Springer Series in
OPTICAL SCIENCES 136

founded by H.K.V. Lotsch

Editor-in-Chief: W. T. Rhodes, Atlanta

Editorial Board: A. Adibi, Atlanta
T. Asakura, Sapporo
T. W. Hänsch, Garching
T. Kamiya, Tokyo
F. Krausz, Garching
B. Monemar, Linköping
H. Venghaus, Berlin
H. Weber, Berlin
H. Weinfurter, München

Springer Series in
OPTICAL SCIENCES

The Springer Series in Optical Sciences, under the leadership of Editor-in-Chief *William T. Rhodes*, Georgia Institute of Technology, USA, provides an expanding selection of research monographs in all major areas of optics: lasers and quantum optics, ultrafast phenomena, optical spectroscopy techniques, optoelectronics, quantum information, information optics, applied laser technology, industrial applications, and other topics of contemporary interest.

With this broad coverage of topics, the series is of use to all research scientists and engineers who need up-to-date reference books.

The editors encourage prospective authors to correspond with them in advance of submitting a manuscript. Submission of manuscripts should be made to the Editor-in-Chief or one of the Editors. See also www.springer.com/series/624

Kai-Erik Peiponen

Risto Myllylä

Alexander V. Priezzhev

Optical Measurement Techniques

Innovations for Industry
and the Life Sciences

With 79 Figures

 Springer

Professor Dr. Kai-Erik Peiponen

University of Joensuu, Department of Physics and Mathematics
P.O.Box 111, 80101 Joensuu, Finland
E-mail: kai.peiponen@joensuu.fi

Professor Dr. Risto Myllylä

University of Oulu, Department of Electrical and Information Engineering
P.O.Box 4500, 90014 Oulun Yliopisto, Finland

Professor Dr. Alexander V. Priezzhev

Lomonosov Moscow State University, Department of Physics and International Laser Center
Vorobiovy Gory, 119992 Moscow, Russia
E-mail: avp2@mail.ru

Springer Series in Optical Sciences   ISSN 0342-4111   e-ISSN 1556-1534

ISBN 978-3-540-71926-7   ISBN 978-3-540-71927-4 (eBook)

Library of Congress Control Number: 2008938195

Typesetting by the authors and SPi, using a Springer LATEX macro
Cover concept: eStudio Calamar Steinen
Cover production: WMX Design GmbH, Heidelberg

SPIN: 11373407   57/3180/spi
Printed on acid-free paper

9 8 7 6 5 4 3 2 1

springer.com

Preface

Optical measurement technique is a wide field because various optical phenomena and properties of light can be used to get information from an object that one wants to get information about. Optical phenomena usually involve light interaction with a medium. The interaction may be linear and involve absorption, dispersion, fluorescence or scattering of light. Properties of light such as amplitude, phase, polarization, wavelength and velocity in a medium provide a rich world to get information from an object. In the case of nonlinear optical phenomena the strong amplitude of the light is the key factor.

In this book we give a view of optical measurement techniques, from the perspective of the authors of this book, especially on applications in industry and life sciences. We are aware about the vast field of the topic. Therefore, we have not intended to cover all possible topics but focus on the following themes, which are applied optical spectroscopy, machine vision, laser velocimetry and measurement of surface quality, position, distance, and displacement.

As concerns industrial optical measurements the newest scientific inventions usually come rather late to practical use. Traditional methods for measurement are favored. Nevertheless, the traditional optical measurement techniques for industry have experienced a renaissance due to strong development of light sources, light detectors, novel type optical elements and CCD cameras.

The field of life sciences is rather wide. Especially the progress of nanotechnology has opened new fields such as nanomedicine and optical tomography. There will be an ever increasing demand to develop optical measurement and imaging technology for life science. In this book we report on some optical measurement techniques that are popular in life sciences.

R M wishes to thank Academy of Finland for financial support during the course of writing this book. K-E P is grateful to PhD Mikko Juuti for technical assistance and data he provided for this book. A V P acknowledges fruitful discussions with colleagues in the International Laser Center of Lomonosov Moscow State University.

Joensuu, Oulu and Moscow
October 2008

Kai-Erik Peiponen
Risto Myllylä
Alexander V. Priezzhev

Contents

1

Introduction

Optical measurement techniques have many advantages over the others if either industrial measurements or, clinical studies in the field of life sciences are considered. These advantages include, noncontact, nondestructive, fast and harsh environment adaptive measurement techniques. Certainly, there are tasks where traditional electronic-based sensors work well and such techniques with optical ones need not be replaced. However, optical measurement techniques have great importance in cases when optical fiber is the only way to reach the object to be measured, or noncontact of the gauge is a must. Typically the industrial circles want to get everything at the lowest price. Usually this leads to compromises in realization of a measurement system. Nevertheless, once the advantages of the optical measurement techniques become clear - for instance to product managers and quality inspection engineers - the barriers for new gauges become much lower. Usually more than speaking with the industrial people about the science behind the optical measurement techniques, it would be more correct to tell them how their problems are solved and how accurate and reliable the device would be. For a scientist, finding a common language between the partners is usually a long jump from academic circles to factories.

The partners presenting life sciences, e.g., medical doctors adapt quite quickly to new optical measurement devices and understand the basic scientific backgrounds. The trend of using novel optical sensors and also imaging techniques will enhance in life sciences. The price is usually not as big an issue as the benefit of the device, when it comes to optical measurement techniques for life science.

In this book we give the readers, insights into some optical measurement technologies that have found applications in various sectors of industry and also life science. Further, we discuss the future trends of optical measurement techniques.

2

Applied Optical Spectroscopy

Optical spectroscopy (a highly accountable reference on applied spectroscopy [1]), has for a long time been a basic tool for materials research and inspection. Optoelectronic innovations such as light sources, detectors, displays, etc. have their foundations in optical spectroscopy. There are different ways to specify optical spectroscopies such as transmission, reflection, fluorescence, spectroscopies. Quite often the spectral region defines the name of the spectroscopy such as UV, visible and NIR spectroscopy. The wavelength of electromagnetic radiation has a great role in the sense that it is sensitive to the basic units of the medium, namely electrons, atoms and molecules that experience the electromagnetic interaction. Nowadays, a rough way to separate different optical spectroscopies into two main classes is to speak of linear- and nonlinear optical spectroscopies. Well established protocols for the use of spectral devices, involving measurements in the field of linear optical spectroscopy, have been furnished, and spectrophotometers can be found in well-equipped laboratories, e.g., for medical and industrial inspection of samples, respectively. The number of industrial products or clinical samples that are inspected by spectral devices is very wide. One basic property is the colour of the sample.

Thanks to the cheaper optoelectronics, robust measurement systems have been built for routine inspection of solid, liquid, and gaseous samples. A variety of commercial miniature spectrometers with software for data processing and displaying are on the markets, and one can easily find up-to-date information of commercial spectrophotometers in the Internet. These miniature spectrophotometers are currently developed so that they offer relatively high resolution of wavelength and concentration of species for various types of real time measurements. Some of these spectrometers involve a fiber optic probe, for instance, for the detection of fluorescence from a product or monitoring fluorescent traces in turbid liquid media. The applications of optical spectroscopy are not limited to medical or industrial purposes only. Security of human beings is becoming more important with globalization. Therefore, spectrometers that can record lethal media such as bacteria and explosives have already been developed and are on their way to general use both in civil and military environments. In forensic studies spectrometers are already everyday life. Future trend in realizing miniature spectrometers for process industry relies on micro electro mechanical systems (MEMS) produced by micro-lithography. Low price, durability and resistance against thermal and mechanical disturbances of such micro-spectrometers will enhance their application in hostile environments of process industry.

Nonlinear optical spectroscopy has been developing fast after the invention and development of high power lasers. Unfortunately, lasers and optical components for nonlinear optical spectroscopy have been relatively expensive, so far. Furthermore, due to complicated systems for materials inspection, and also weakness of the signal nonlinear optical spectroscopy, as far as we know, they have been able to evince great interest in the circles of process industry. One reason may also be that the theoretical background of nonlinear optical spectroscopy is rather difficult to understand for practioners, and may therefore hinder the adoption of such a technology, although it might solve many problems that appear, e.g., in adjusting process parameters. Fortunately, due to progress of powerful small size solid state lasers that can be purchased, nonlinear optical spectroscopy has a potential of becoming much stronger in the field of applied spectroscopy, especially biomedical optics.

In this chapter, we deal with typical optical spectroscopies that are currently used or are developed in the industry and research laboratories, and deal briefly the case of two-photon induced fluorescence that belongs to the field of nonlinear optics. We omit in this book the important sub-field of applied spectroscopy namely Fourier transform spectroscopy. There is a recent publication on this subject [2], which is highly recommended as a standard source for the physical backgrounds and applications of this sub-field. Here we just mention that Fourier transform infrared spectroscopy (FTIR) has various applications in analysis and identification of liquids, gases, and mixture of gases. Portable Fourier transform spectrometers have libraries in the memory of the computer that enable identification of different gas components. This has importance, in remote sensing of air pollution of exhaust from chimneys of factories and power plants, and also traffic. There are also military applications related to the FTIR technology.

2.1 Transmission Spectroscopy

Measurement of wavelength-dependent light transmission of a collimated light beam through transparent and colored planar solid objects is a routine procedure in materials inspection. Nowadays robust portable spectrophotometers for transmission measurement are commercially available. In many applications the spectrophotometer is assumed to monitor transmittance of a product. The transmittance is defined by the expression

$$T(\lambda) = \frac{I(\lambda)}{I_o(\lambda)}, \tag{2.1}$$

where I_o is the incident light intensity, I the transmitted light intensity upon the detecting area of the detector and λ the wavelength of the light. Using transmittance it is possible to gain information on the local thickness (d) of the product, which is assumed to be homogenous, via the Beer–Lambert law

$$I = I_o \exp(-\mu(\lambda)d), \tag{2.2}$$

where μ is a material parameter, i.e., the wavelength-dependent absorption coefficient of the medium. It is worth emphasizing that the absorption coefficient of a medium depends not only on the wavelength of the incident light but also on the thermodynamic condition of the sample. The temperature dependence of the object has to be specially taken into account in the measurement of a temperature sensitive sample. Typically there may appear fluctuation of the sample temperature, which is usually due to external heat sources, for instance, those present in industrial environments. Temperature-regulated spectrometers are recommended for such conditions. Thickness variation of a semitransparent sheet is obtained from (2.2). If we are working with a priori known material then the absorption coefficient has usually been measured in a laboratory at some appropriate spectral range. Hence, one can monitor the thickness change of a product using two or more different wavelengths, obtained by filtering from a white light source with a filter system, and using a two- or multidetector system. An alternative device may involve two or more lasers as collimated light sources, with well-known accurate wavelengths. Many times at industrial site external vibrations appear, for instance, due to the conveyor belt of the product. This means that the product, e.g., a plastic sheet may experience wavy or other type of motion. In order to get "a frozen" probed thickness, the probe beam is usually chopped. In the case of using two probing wavelengths they are usually chosen in a manner that the material absorbs at one wavelength say λ_1, whereas at λ_2 the absorption of the object is negligible. From (2.2) we can then solve the thickness of the object as follows:

$$d = \frac{1}{\mu(\lambda_1)} \ln \frac{I_o(\lambda_2)}{I(\lambda_1)}. \tag{2.3}$$

Detection of the intensity ratio by two detectors is of great importance since the aging of the light source is not a big issue then. If the product is a sheet that is moving on a conveyor belt and the sensing head has a fixed position, then thickness of the product along a line can be assessed. If a scanning measurement head is traversing across the machine direction, thickness of the products is obtained along a "Z-shaped" path. Obviously, one cannot probe the whole product with this kind of gauge, but usually this is not a big issue. Statistical analysis of the thickness and product history are usually stored into the memory of a computer so that properties of the product can be checked many years after its manufacture. In general, such storage of production history of any product is important. This is true particularly in the case of fracture of the purchased product because then it is usually possible to track production parameters of the broken product to check for responsibility of damage.

Transmission measurement by the two detector system is useful, not only in production environments of new products, but also in sorting of recycled plastic waste [3].

Basic metal industry that produces cold-rolled metal, which may take the form of a sheet or a roll, sells these products, for instance, to the automotive industry. Usually there is a requirement to protect the metal surface against corrosion, especially if the restoration time of the metal is relatively long. Thus from the customer side it is necessary to protect the metal sheet or a roll by a thin oil layer. It is important to optimize thickness of the oil layer since it has usually to be removed, before further processing, by using washing agents. For the sake of optimization of the thickness of the oil layer, it is measured at the metal production site; thickness of the layer can be checked by the customers also at their production sites. Optimization is important not only for proper protection against corrosion but also for reducing the amount of washing agents. The latter is of crucial importance for the purpose of decreasing pollution of the environments. The measurement technique may exploit, for instance, the absorption of the IR radiation in the protection oil, where absorbing units are hydrocarbons. Since the oil layer covers opaque metal, the measurement is based on transflectance, i.e., the oblique incident radiation is transmitted through the oil layer but reflected from the metal substrate. Analysis of the oil film thickness can be based on the utilization of (2.1). There are commercial portable devices available for monitoring oil film thickness. Reliability of the measurement result depends to some extent on the surface roughness of the metal product under the oil film. In Fig. 2.1 we show a principle of the oil film hand held measurement system of the transflectance on a spot of a metal roll.

Simultaneous measurement of film thicknesses and refractive indices of several solid layers is possible using the principles of inteference and reflection in thin films by scanning a spectrum from UV- to NIR region. The thickness, absorption and refractive index of a film on a substrate can be obtained using also an ellipsometer. Devices based on above mentioned principles are already in the markets.

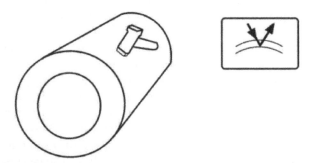

Fig. 2.1. Principle of oil film measurement using IR- absorption of hydrocarbon groups (*right*). An application is detection of oil surface density from steel sheet and roll (*left*)

Quite often we have to measure the transmission of liquids such as water where different constituents are mixed. In such a case we exploit a cuvette with well-known thickness such as 1 mm or 1 cm depending on the optical density of the sample. Then the transparency or colour of the liquid can be estimated with the aid of the transmission spectrum using the transmittance T. Real refractive index (n) and its change (Δn) are other quality parameters of the light absorbing liquid, which provide information about the concentration of the liquid. In the case of transmission measurement technique we can calculate, in addition to the absorption coefficient of the liquid, refractive index change as a function of the wavelength from measured data. From (2.2), one can solve the absorption coefficient, and then get the extinction coefficient (k) of the liquid, which is given by the expression

$$k(\lambda) = \frac{\mu(\lambda)\lambda}{4\pi}. \tag{2.4}$$

The real refractive index change and extinction coefficient are coupled together by Kramers–Kronig relations, and also by their modifications [4]. A useful method, especially when using band-limited transmission data only, is to extract the frequency-dependent real refractive index change from a so-called singly subtractive Kramers–Kronig (SSKK) relation [4], and thereafter perform a cross checking of the calculated data by the conjugated SSKK relation. The two SSKK relations require an anchor point where optical constant is a priori known. The information at the anchor point can usually be measured by other means, e.g., in the case of liquids, by a reflectometer. The SSKK relations are as follows:

$$n(\omega') - n(\omega_1) = \frac{2(\omega'^2 - \omega_1^2)}{\pi} P \int_0^\infty \frac{\omega k(\omega)}{(\omega^2 - \omega'^2)(\omega^2 - \omega_1^2)} d\omega, \tag{2.5}$$

$$\frac{k(\omega')}{\omega'} - \frac{k(\omega_1)}{\omega_1} = \frac{2(\omega'^2 - \omega_1^2)}{\pi} P \int_0^\infty \frac{n(\omega) - n_\infty}{(\omega^2 - \omega'^2)(\omega^2 - \omega_1^2)} d\omega, \tag{2.6}$$

where ω_1 is the anchor point frequency, P denotes Cauchy principal value and n_∞ is the high energy value of the refractive index. A code of a numerical algorithm for the SSKK analysis is presented in the book of Lucarini et al. [4].

For example in Fig. 2.2, we show the real refractive index change and extinction coefficient of 12 different commercial red wine products. First the transmittance of the red wine samples is measured using 1 mm cuvette at room temperature. The extinction coefficient is calculated with the aid of (2.4), and the refractive index change is calculated using (2.5). One application of the complex refractive index data is to help wine producers in research and development of red and other wines. Another application is to test the authenticity of purchased wine, in other words to check that the bottle and its etique match with the contents. This is to prevent counterfeit of wine and other drinks.

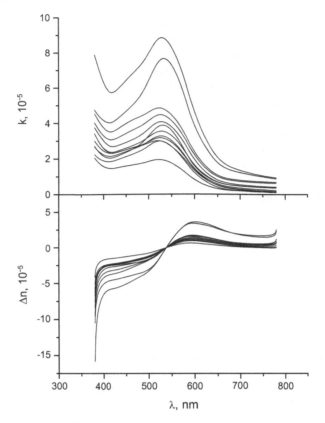

Fig. 2.2. Extinction coefficient (*upper panel*), and real refractive index change (*lower panel*) of some commercial red wines

In Fig. 2.2 one can identify the anthocyanins in the vicinity of 520 nm due to strong absorption of light. This is important in the color formation of red wines [5]. Color is considered as an important factor of red wine quality. Color coordinates (CIE Lab) of red wine can be calculated from the transmission spectrum. Another quality factor of red wine is the concentration of tannins which can be recognized due to strong light absorption at the vicinity of 280 nm.

Transmittance data of wine products can be utilized in quality inspection during and after the fermentation process. Fiber optic sensors provide means for in situ optical inspection of optical properties of wines [6]. Naturally, transmission data can be utilized in screening of whiskies and other colored alcohol drinks.

A spectral analysis reveals that the set of 12 red wines differ from each other. Hence studying spectral features and the color of the red wine one can gain information, e.g., about the authenticity of the wine. This is an important method in preventing counterfeiting of red wines and other liquid products.

The absolute value of the refractive index of red wine provides information about the density of the wine which depends on alcohol and sugar contents. Absolute value is obtained usually by a refractometer. We will describe the measurement of refractive index of liquids in Sect. 2.3.

Since red wines are more or less turbid liquids, scattering of light has also a role in the detection of true transmittance. By choosing a thin cuvette this light scattering can be satisfactory eliminated. Measurement of turbidity of wines and other light scattering liquids is one indicator of the quality of such liquids. In the next section we deal with the issue of turbidity in more detail.

2.2 Measurement of Turbidity of Liquids

The light scattering theory of Mie [7] is the basis of light interaction with spherical scattering centers, whose size matches the wavelength of the incident light, in gas or liquid phase. The theory of Mie works well in such situations where mono- or polydispersive spherical scatterers, with well known complex refractive index, occupy a relatively low volume fraction in the host medium. If the shape of the scatterer is complex, Mie theory becomes invalid. Furthermore, if the complex refractive indices of different types of scatterers, appearing simultaneously in the scattering volume of probe light is not known, one usually faces problems in rigorous interpretation of the measured signal. However, if we do not want to know specific properties of the scatterers but merely wish to gain information on the turbidity of the liquid, i.e., reduction of transparency of a liquid caused by undissolved matter, simple measurement geometry can be exploited. A modern device for the measurement of turbidity is called a nephelometer. The idea behind this device is to measure light scattering at 90° scattering geometry as shown in Fig. 2.3. Although the signal is a nonlinear function of turbidity, this geometry is sensitive to light scattering from particles.

The unit turbidity of this device is nephelometric turbidity unit (NTU) that may range between 0 and 10 000 NTU. Calibration is performed against

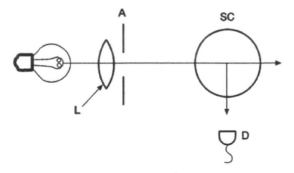

Fig. 2.3. Schematic diagram of a nephelometer. L, lens; A, aperture; SC, sample cell; and D, detector

formazin standard. It is obvious from Fig. 2.3 that the detection of scattered light depends not only on the concentration of the scatterers (multiplescattering) but also on the absorption of light by the scatterers and the liquid host, as well as on the spatial refractive index changes of the scattering medium. Evidently, it is difficult to derive a general theory for analysis of the signal. Thus we usually simply make use of the intensity of the light detected at the 90° scattering angle. Naturally, other scattering angles can be utilized in the assessment of turbidity in a wider scope as it has been shown for port wine [8] by recording turbidity of the port wine, during the production process, as a function of wavelength. Multifunction commercial sensors for monitoring quality of process waters in industry, or wastewater in treatment plants, are commercially available and include also the mode of gaining the turbidity. Process monitoring robust devices have been designed and produced, for measurement of consistency of pulp solutions in paper industry. The problem of the coloration of the turbid sample is usually avoided by choosing the operation wavelength at infrared region. In the event of strong turbidity, the backscattering measurement mode of the turbidity sensor is favorable. In the same measurement head forward- and backscattering measurement modes can be included in order to cover wide dynamical range of the turbidity. The common problem with an optical sensor head, when monitoring liquids of biological or industrial origin, is the contamination of the probe window. One solution to avoid this is to implement ultra sonic cleaning unit in the sensor head. Development of coating technology for noncontaminating surfaces is in strong progress, and the trend is to utilize nanostructures as over layers of the product, such as a wide variety of window materials.

In Table 2.1 we show NTU readings, in addition to the refractive index, of the 12 commercial red wines. Obviously, turbidity is subject to stronger fluctuation than the refractive index of these red wine samples.

Table 2.1. Production country, refractive index, alcohol volume % and turbidity of a set of 12 commercial red wine samples

Sample	Country	Refractive index (ABBE)	Alcohol %	Turbidity (NTU)
1	Italy	1.3440	12	5.6
2	Spain	1.3444	12.5	29
3	Spain	1.3441	12.5	24
4	France	1.3437	13	20
5	Chile	1.3462	14	190
6	France	1.3451	13	24
7	South Africa	1.3452	13.5	47
8	France	1.3451	12	10
9	Portugal	1.3439	12.5	54
10	USA	1.3448	13	43
11	France	1.3442	13.5	9.4
12	Argentina	1.3452	14	64

2.3 Reflection Spectroscopy

In this section we deal with the reflection measurement mode that has practical utility both in inspection of liquids and porous media. In the case of liquids a probe window can be introduced for on-line operation at industrial site in order to monitor the process parameters of the liquid condition. The measurement system may yield feedback to a process- or central computer for the purpose of optimizing and controlling the process parameters. In the case of porous media the inspection usually happens at the laboratory of the industrial site; depending on the object, on-line devices can also be furnished.

2.3.1 Refractometer

In the context of liquids, their refractive index is a basic optical quantity that is used in quality inspection. By measurement of the refractive index we can identify different liquids, get information on concentration of sugar, salt, proteins, acids, etc. diluted in water, alcohol in water, lactose in milk and so on. The principles of light reflection and refraction have found applications, especially, in inspection of refractive index of transparent liquids products and also process liquids of industrial processes. A classical refractometer is based on the Abbe refractometer. This device makes use of total reflection of light, as follows:

$$n = \sin \theta_c, \tag{2.7}$$

Where n is the relative refractive index, i.e., the ratio $n = n_{\text{liquid}}/n_{\text{prism}}$ and θ_c is the critical angle of total reflection. The refractive index of the prism is a priori known. Hence by measurement of the critical angle, one gets the refractive index of the liquid. Dispersion data of typical prism materials can be found, for instance, from catalogues of companies that sell optical elements.

There are various types of commercial refractometers both for laboratory and industrial environments. Typically they measure refractive index of juice, beverages, soft drinks, wines and beer, and milk products in the food stuff industry. Naturally they have applications also in the fields of chemical, petro-chemical and pharmaceutical industries. In the construction of a refractometer a LED or a white light source can be used as a light source. Using a filter some specific wavelength from a relatively wide spectral range light source, usually corresponding emission line of sodium in the vicinity of 589 nm, is chosen for monitoring of the real refractive index of the liquid to be inspected. It is possible to also use a laser as a light source of the refractometer. Aging of the light source is not a critical factor since the idea is to find the border of abrupt change of the intensity of the reflected light at the critical angle of incidence. The signal is detected by a linear photodiode array or a CCD-camera. In environments where the liquid is hot the sensor head is cooled by water circulation. For measurement of the refractive index of high pressure liquid the mechanical resistance of the sensor head is an issue. Wear of the probe window, in monitoring high pressure liquid flow, can be minimized by choosing sapphire as the prism material. There are multihead configurations for in-line monitoring of process liquids in piping systems of process industries. Accuracy of the measurement of the refractive index is about 10^{-4}.

In the event of light absorbing and/or turbid liquids the concept of total reflection becomes questionable. Indeed, the classical Abbe device is based on the abrupt jump of the reflected light intensity at the critical angle of reflection. For instance, for colored liquids there is no such abrupt jump of the reflected light intensity for a fixed wavelength of the incident light. It means that we cannot locate the critical angle. This is a problem that can be overcome by recording the reflected light intensity as a function of the angle of light incidence, and forming the second derivative from the reflectance. The maximum of the derivative gives the location of the apparent critical angle. Color and turbidity of the liquid is no more a great issue in design and construction of modern process refractometers. The refractive index interval obtained by a process refractometer depends on the choice of the prism material and the focusing geometry of the light that is incident on the prism–liquid interface. For instance, one can construct a device where a focused laser beam provides an angle distribution for the measurement of relatively small changes of the refractive index of the liquid. From the Internet one can check rather easily the availability of commercial refractometers.

2.3.2 Reflectometer with Wavelength Scanning Mode

Estimate on complex refractive index of opaque media, such as slurries, at wide spectral range can be obtained using a reflectometer and relevant spectra analysis methods [9]. An alternative apparatus is an ellipsometer, which provides information on the complex refractive index of solid sample but usually for a relatively narrow spectral range. The principle of ellipsometry can be utilized for monitoring purposes in process industry, but then one usually makes use of a laser light source. Fig. 2.4 shows a multifunction reflectometer, which was developed for the purpose of process water analysis in pulp and paper mills. One can chose different measurement modes with this prism reflectometer. These include scanning of the angle of incidence or the wavelength. In addition, linear polarization, such as s- and p-polarizations of the incident light can be chosen.

The spectra analysis is most convenient by using the Fresnel's formulas for reflectance of s- or p-polarized light as follows:

$$R_s(\omega) = \left| \frac{\cos\theta - \sqrt{N^2(\omega) - \sin^2\theta}}{\cos\theta + \sqrt{N^2(\omega) - \sin^2\theta}} \right|^2, \tag{2.8}$$

and

$$R_p(\omega) = \left| \frac{N^2(\omega)\cos\theta - \sqrt{N^2(\omega) - \sin^2\theta}}{N^2(\omega)\cos\theta + \sqrt{N^2(\omega) - \sin^2\theta}} \right|^2, \tag{2.9}$$

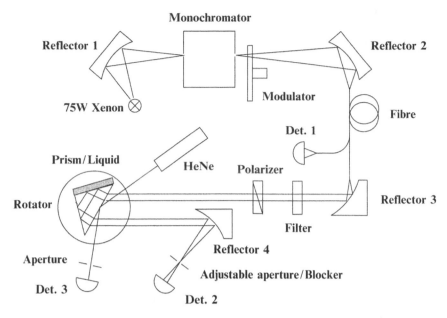

Fig. 2.4. Schematic diagram of a reflectometer

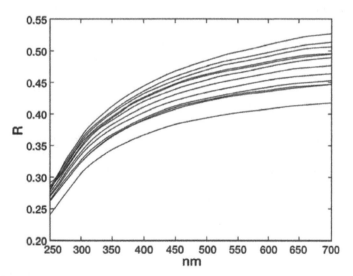

Fig. 2.5. Reflectance from 12 red wine samples as a function of wavelength. *Curves* were recorded by a reflectometer using *s*-polarized light

where θ is the angle of incidence, and N is the relative complex refractive index. The simplest case, in the frame of (2.8) and (2.9), occurs when the extinction coefficient of the liquid is so small that it can be neglected. In Fig. 2.5 we show reflectance for the 12 red wines that we have considered

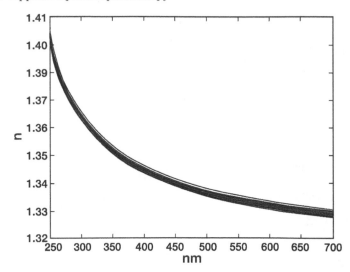

Fig. 2.6. Real refractive index of 12 red wines obtained using data of Fig. 2.5

previously. Data is recorded as a function of wavelength for a fixed angle of incidence and using s-polarized light. Under the assumption of relatively low extinction coefficient one can solve the refractive index of the red wines from (2.8), since there is only one unknown quantity namely the real refractive index n. Dispersion curves for the 12 red wines, using data of Fig. 2.5, are shown in Fig. 2.6. From Fig. 2.6 we observe that a relatively good way of monitoring concentration of the red wines can be found by integrating the refractive index curves. Use of the sodium D-line gives poorer sensitivity to some extent to distinguish differences between these wine samples.

In the case of scanning the angle of incidence for a fixed wavelength, an estimate of the complex refractive index can be obtained using an optimization method [9]. In this method one minimizes the least square sum of the difference between the reflectance obtained from the theory, (2.8) or (2.9), by inserting candidates for the complex refractive index N as follows:

$$S = \text{Min} \sum_{\theta} [R_{\text{m}}(\theta) - R_{\text{t}}(\theta)]^2, \qquad (2.10)$$

where R_{m} and R_{t} are the measured and theoretical reflectance, respectively.

Reflection coefficients of the medium, corresponding to the reflectance (2.8) and (2.9), are complex numbers due to the complex refractive index. Reflection coefficients for s- and p-polarized light can also be given in the following polar forms

$$r_s(\omega) = |r_s(\omega)|\, e^{i\varphi_s(\omega)}$$
$$r_p(\omega) = |r_p(\omega)|\, e^{i\varphi_p(\omega)} \qquad (2.11)$$

using the concepts of amplitude and phase of a complex number. In the wavelength scanning mode and fixed angle of incidence, an estimate of the spectral dependence of the complex refractive index can be obtained by phase retrieval procedure using either singly or multiply subtractive Kramers–Kronig relations or by maximum entropy method (MEM) [4,10]. However, when utilizing SSKK one has to take care, especially when measuring reflection spectra, of the oblique incidence of p-polarized light [11]. In this particular case when the condition

$$n_\infty \leq \tan \theta \leq n_{\text{static}} \tag{2.12}$$

For a liquid (solids as well) is valid, K–K relations give an erroneous phase for the complex reflectivity. It means also that one gets erroneous complex refractive index of the liquid which can be solved using (2.9) and the latter form in (2.11). If the above condition (2.12) is not valid for p-polarized oblique incident light, or if one uses oblique incident s-polarized light, or the measurement happens at normal incidence then one can exploit the SSKK relations below, with one anchor point ω_1, for the phase retrieval and cross checking of calculated data

$$\ln |r(\omega')| - \ln |r(\omega_1)| = \frac{2(\omega'^2 - \omega_1^2)}{\pi} P \int_0^\infty \frac{\omega\phi(\omega)}{(\omega^2 - \omega'^2)(\omega^2 - \omega_1^2)} d\omega, \tag{2.13}$$

and

$$\frac{\varphi(\omega)}{\omega'} - \frac{\varphi(\omega_1)}{\omega_1} = -\frac{2(\omega'^2 - \omega_1^2)}{\pi} P \int_0^\infty \frac{\ln |r(\omega)|}{(\omega^2 - \omega'^2)(\omega^2 - \omega_1^2)} d\omega. \tag{2.14}$$

The reflectance and the reflection coefficient are coupled by the well-known relation $R = |r|^2$. Note that in practical analysis one has to limit the integration inside a finite spectral range. Such a procedure is a source of error. This error is reduced by the anchor point technique. The reason for the error decreasing is the better convergence of the SSKK relations than conventional Kramers–Kronig ones

Next we spend some time with phase retrieval using MEM, which may not be familiar for the readers. MEM can be used instead of K–K analysis for any situation, including the condition (2.12). Hence MEM is a more general method than K–K analysis, although the basis of MEM is not available in physics. MEM merely presents a mathematical method from information theory. Below we describe some features of MEM. In this method a reflectance spectrum, measured at a finite angular frequency range, is compressed into the interval between 0 and 1 by change of variable

$$\nu = \frac{\omega - \omega_{\text{start}}}{\omega_{\text{end}} - \omega_{\text{start}}}. \tag{2.15}$$

The complex reflection coefficient is obtained from a series expansion

$$r(\nu) \cong \frac{|d_o| e^{i\phi(\nu)}}{\left| \sum_{m=0}^M d_m \exp(-2\pi i m \nu) \right|}, \tag{2.16}$$

where the coefficients d_m are obtained from a set of Yule–Walker equations

$$\sum_{m=1}^{M} d_m C(m-p) = \begin{array}{l} |d_o|^2\,, m = 0 \\ 0, m = 1, \ldots, M, \end{array} \tag{2.17}$$

and where autocorrelations C are obtained from the integral

$$C(\nu) = \int_0^1 |r(\nu)|^2 \exp{[i2\pi q\nu]}\, d\nu. \tag{2.18}$$

In the case of MEM we use data on the measured wavelength range only. That is to say no data extrapolations are performed. Since (2.16) is an approximation we usually have to correct the phase angle of the complex reflection coefficient. For that purpose we must have phase information at anchor points, which are located inside the measurement range. Usually two anchor points are enough to get a good estimate for the complex reflection coefficient, and therefore also for the complex refractive index of the medium, which is in liquid or solid phase. Angle correction is carried out using so-called "error phase" ϕ, which typically is a smooth nearly linear function with slow variation. The error phase can be presented as a polynomial interpolation

$$\phi(\nu) = \sum_{r=0}^{R} B_r \nu^r, \tag{2.19}$$

where the coefficients are obtained from a Vandermonde system

$$\begin{pmatrix} 1 & \nu_o & \cdot & \cdot & \nu_o^L \\ 1 & \cdot & \cdot & \cdot & \nu_1^L \\ \cdot & \cdot & & & \cdot \\ \cdot & \cdot & & & \cdot \\ 1 & \nu_L & \cdot & \cdot & \nu_L^L \end{pmatrix} \begin{pmatrix} B_o \\ \cdot \\ \cdot \\ \cdot \\ B_L \end{pmatrix} = \begin{pmatrix} \phi(\nu_o) \\ \cdot \\ \cdot \\ \cdot \\ \phi(\nu_L) \end{pmatrix}. \tag{2.20}$$

In order to increase the linearity of the error phase another compression of the spectrum into a narrower spectral range is performed using the following data fitting procedure

$$|r_p(\nu)|^2 = \begin{array}{l} |r(\nu)|^2\,, 0 \leq \nu < w_K(\omega_1) \\ |r(\nu)|^2\,, w_K(\omega_1) \leq \nu \leq w_K(\omega_2) \\ |r(\nu)|^2\,, w_K(\omega_2) < \nu \leq 1, \end{array} \tag{2.21}$$

where

$$w_K(\omega) = \frac{1}{2K+1}\left(\frac{\omega - \omega_1}{\omega_2 - \omega_1} + K\right), \tag{2.22}$$

and

$$\nu = \frac{w_K(\omega) - w_K(\omega_1)}{w_K(\omega_2) - w_K(\omega_1)},\tag{2.23}$$

where K is a positive integer.

A crucial difference between K–K analysis and MEM is that in the case of the former we deal with logarithm of reflectivity, whereas with MEM we deal with the reflectivity itself. Hence, we can avoid the singularity which appears with logarithm as the complex abscissa is equal to zero.

In Fig. 2.7 we show reflectance and complex refractive index of a water–lignin solution, which is present in the process of pulping of wood for paper production. The water–lignin solution is optically very dense, which

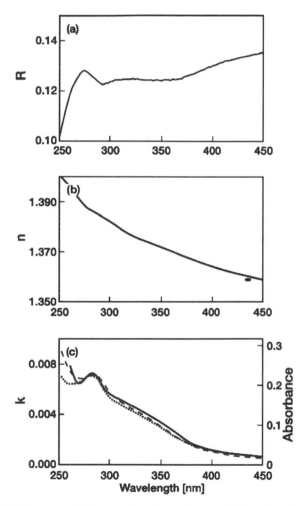

Fig. 2.7. (a) Reflectance, (b) real refractive index, and (c) extinction coefficient of a water–lignin solution

means that its intrinsic complex refractive index can be extracted only by measurement of the reflectance. The complex refractive index is extracted from the reflectance spectrum with the aid of the MEM analysis described above.

Due to scattering particles, which may be present in the liquid, the probe window of the prism is in many cases subject to contamination as a function of time. Sometimes it is possible to assess the correct reflectance by modeling the contamination as it was shown for pulping liquids [12, 13]. Fortunately the contamination layer can usually be removed by ultra sonic washing of the probe window.

After the somewhat lengthy theoretical treatment above we conclude this section by noting that the basic concept of the reflectometer (Fig. 2.4) provides a device for a multi-measurement scheme. It is possible to utilize a Dove prism that allows measurement of transmission, reflection and scattering of light from liquid samples with low or high turbidity. In addition rotation of light polarization can be used for measurement of optical activity of the liquid. With the reflectometer one gets information on the complex refractive index of extremely difficult objects such as offset inks used in printing houses [14, 15] and also birefringence of pigments used in paper industry [16]. Advantage with this kind of multifunction spectrometer is that one can avoid practical problems such as different practices, calibration, costs, maintenance, time consumption, and space requirements.

Surface Plasmon Resonance Spectrometer

Plasma oscillation in the surface mode can be generated using a prism, with one face coated by a thin metal film, and a laser as a light source. If the complex permittivities of the metal film and liquid sample are chosen properly, a surface plasmon resonance (SPR) can be exited for a fixed wavelength at an angle that is larger than the critical angle of reflection. SPR occurs when the wave number of the incident field parallel to the surface matches with the complex wave number of the surface plasmon. A detailed description of the physics of the SPR can be found from [17].

Typically the thickness of the metal film is ca. 50 nm. In laboratory experiments usually silver film is used due to the relatively strong SPR signal, whereas in commercial sensors gold film has been favored. Commercial SPR sensors make use of the Kretchmann's configuration [18] and a focused light beam impinging on the interface between the sample and the prism [19]. The focused beam automatically provides a range of incidence angles of the light rays; thus there is no need to rotate the probing prism. Obviously, the range of variation of the refractive index is limited by the magnitude of the cone of the incident light and the refractive index of the probe prism. However, usually the range of the variation of the refractive index (variation of concentration) of certain liquid can be estimated and taken into account in the construction of the SPR sensor. A photodetector, array of detectors or a CCD-camera is

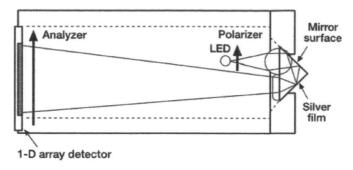

Fig. 2.8. Schematic diagram of SPR sensor

used for the detection of the spatial variation of a dip in the intensity of the reflected light. In Fig. 2.8 we show a schematic diagram of a SPR sensor.

The list of different kind of applications of a SPR sensor is long including, among others, measurement of optical properties and thickness of metal films, adsorption (or desorption) of gas molecules, drug discovery, food stuff applications, loadings of pigments used in paper, - paint and other sectors of industry, where liquids and pigments play important role. In some applications the metal surface has a special feature so that its texture is modified by introducing a lipid bi layer, adsorbed co-polymers, etc. This is important especially in monitoring of dynamic biological interactions where the liquid to be measured is under a flowing process in a flow cell that is coupled into the SPR measurement head [20]. Such a scheme is useful, e.g., in monitoring the kinetics of protein adsorption to biomaterials. A recent trend of SPR sensing is the research of proteins and protein interactions in low-gravity conditions on board a space station [21]. A DNA sensor based on SPR has been realized [22].

Most typical application of an SPR sensor is the measurement of the refractive index of a liquid. Analysis of refractive index of the sample is based on the following formula

$$\omega n_3 \sin\theta_{sp} = \sqrt{\frac{\varepsilon_1\varepsilon_2}{\varepsilon_1+\varepsilon_2}}. \tag{2.24}$$

This formula makes it possible to obtain at one wavelength of a laser, or at a discrete set of wavelengths, the refractive index of liquids by tuning the angle of incidence. Figure 2.9 illustrates reflectance dip curves for three homogenized milk samples. Sample 1 has 0.0004% fat, 3.48 proteins, and 4.96% lactose; sample 2 has 1.53% fat, 3.41 proteins, and 4.87% lactose; and sample 3 has 3.55% fat, 3.57 proteins and 4.74 lactose. All these samples contain water and small amounts of other constituents. After homogenization the size of fat particles is less than one micron and protein particles are even smaller. From Fig. 2.9 we can observe that location of minimum as well as half width of the dip are subject to change as a function of the fat volume percentage of the milk.

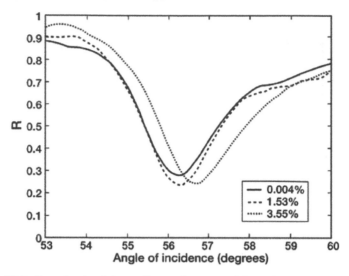

Fig. 2.9. SPR dips obtained for milk as a function of fat volume concentration [23]

Traditionally, the analysis of the reflectance curve in SPR sensing is based on observing resonance dip at the angle θ_{sp}, which shifts to different wavelengths due to the light dispersion of the liquid, metal film and the prism material. It is important that the system is thermally stable during the measurement since small temperature gradient may induce change location of the resonance angle.

In the case of surface plasmon resonance spectroscopy, we can make use of a reflectometer similar to that described in Sect. 2.3.2 but the face of the prism, which is in contact with the liquid, is covered by thin metallic film. Now the angle of incidence is fixed while the wavelength is scanned. While recording of the spectra, various finger prints of different species can be obtained. This is the major advantage of SPR spectroscopy, in addition to sensitivity to small variations in liquid concentration. In Fig. 2.10 are shown SPR dips that were obtained from water for different wavelengths. In the case of absorbing liquids ambiguity may appear in the location of the SPR dip and a single absorption peak [24].

Surface plasmons can be also generated from metallic particles embedded in a dielectric medium. Hence the interaction of the light waves does not depend on the use of a prism as coupler. Quite recently, research on nanostructures has become one of the most important fields in technology and life sciences. One of the most important fields is nanomedicine where new generations of cancer diagnostic and therapeutic means, which will dramatically improve cancer outcomes, are developed with aid of nanotechnology. In other words nanoparticles, such as gold nanoshells, quantum dots, etc. will be used for tissue targeting, sensing and imaging, and also localized therapy. An

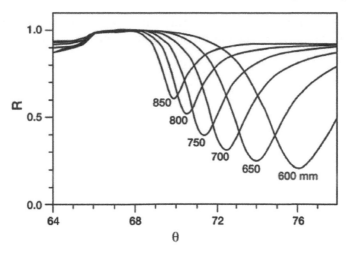

Fig. 2.10. SPR dips for water at different wavelengths. Step at the *curves* in the vicinity of 65° is due to the total reflection

advantage of nanoparticles is the use of lower doses. Gold nanospheres coupled to DNA or protein probes can be used in diagnostics of small amounts of proteins that appear in the case of cancer, cardiovascular disease, Alzheimer disease, etc. The unique property of a gold nanosphere is that one can probe proteins through optical phenomena. Quantum dots are usually nanoscale semiconductors with fluorescing tags. They absorb light at narrow spectral band but emit at long wavelengths at the infrared range. Quantum dots with protein coating can penetrate cells because cells regard them as proteins. Common aspect with nanoparticles is that their optical properties depend on their size, shape, and environment. Localized surface plasmon resonance spectroscopy in medical diagnostics that is based on the use of nanosensors is in progress [25]. In the event of using a large number of nanoparticles in medicine nanotoxicology may be an issue.

The case of several absorption peaks of dielectric nanoparticles, which are in a water matrix, appear simultaneously with the SPR dip was analyzed by Vartiainen et al. [26] from the reflectance data with the aid of the MEM. The extraction of the complex refractive index from SPR data is possible from the equations that are obtained by considering multiple Fresnel reflection in a thin film sandwiched between two media (liquid and prism) and with different complex refractive indices, respectively. The machinery for such an analysis is based on the reflectance

$$R_p(\theta) = \left| \frac{r_{\mathrm{pm}}(\theta) + r_{\mathrm{ml}}(\theta) \exp\left[2i A_z(\theta)d\right]}{1 + r_{\mathrm{pm}}(\theta) r_{\mathrm{ml}}(\theta) \exp\left[2i A_z(\theta)d\right]} \right|^2, \tag{2.25}$$

where r_{pm} is the complex reflection coefficient for the prism–metal interface, r_{ml} is the reflection coefficient at the metal film–liquid interface, d is the thickness of the metal film and A_z is the scalar component of the wave vector in the direction of the normal of the metal film. Furthermore,

$$r_{\mathrm{pm}} = \frac{\dfrac{A_{z,\mathrm{prism}}}{\varepsilon_{\mathrm{prism,r}}} - \dfrac{A_{zm}}{\varepsilon_{mr}}}{\dfrac{A_{z,\mathrm{prism}}}{\varepsilon_{\mathrm{prism,r}}} + \dfrac{A_{zm}}{\varepsilon_{mr}}}, \tag{2.26}$$

$$r_{\mathrm{ml}} = \frac{\dfrac{A_{zm}}{\varepsilon_{mr}} - \dfrac{A_{z,\mathrm{liq}}}{\varepsilon_{\mathrm{liq,r}}}}{\dfrac{A_{zm}}{\varepsilon_{mr}} + \dfrac{A_{z,\mathrm{liq}}}{\varepsilon_{\mathrm{liq,r}}}}, \tag{2.27}$$

where $\varepsilon_{\mathrm{prism,\,r}}$ is the relative permittivity of the prism, ε_{mr} the relative complex permittivity of the metal film (usually the bulk permittivity is used in data analysis) and $\varepsilon_{\mathrm{liq,r}}$ the corresponding relative permittivity of the liquid (or gas). The wave number is given by the expression

$$A_{zj} = \left[\varepsilon_{jr} \left(\frac{\omega}{c} \right)^2 - A_x^2 \right]^{1/2}, \tag{2.28}$$

where

$$A_x = n_{\mathrm{prism}} \frac{\omega}{c} \sin\,\theta. \tag{2.29}$$

Before closing this section we emphasize that SPR data can also be exploited in the detection of thickness of the metal film on the prism face. This is important not only in basic research but applications where thickness of the metal film is subject to wear. For instance, wear can be due to laminar or turbid flow of solid particles that pass the probe window, or due to chemical corrosion. Monitoring of the metal film allows better calibration of the SPR gauge.

2.4 Measurement of Diffuse Reflection from Porous Media

Light interaction with porous media is complicated. Complications originate from multiple scattering of light (Mie scattering theory is invalid), random location, and different geometries and complex refractive indices of the particles that constitute the porous media. A simple example of such a porous medium is paper. Paper has a fiber network, fillers, fines, and pigments. Due to porosity of a medium, diffuse transmission and/or reflection of light appear. The diffuse light provides spectral data of the sample. One can use a photogoniometer for detection of the scattered light but such a device is tedious to use and rather expensive. Normally diffuse light is detected using a spectrophotometer, which includes an integrating sphere. The angle of light incidence is usually fixed near the normal incidence. It is possible to also measure specular component of light with the integrating sphere. Portable spectrophotometers with white light sources incorporated by an integrating sphere are available, and such spectrophotometers can also be utilized for detection of the colour of the object.

In the event of thick samples, it is reasonable to use the diffuse reflection measurement mode. Examples of thick and light diffusing objects are, pharmaceutical tablets, pile of papers, ceramics and so on, which can be inspected using the diffuse light. Here we concentrate on the inspection of paper in some depth. Opacity is a crucial parameter of paper, especially for paper grades that are produced for printing purpose. Thus opacity is important in characterizing the quality of news papers, magazines, and books. High opacity means that one cannot distinguish printing on the opposite side of the paper. Obviously, high opacity (=high nontransparency) is a measure of the quality of paper for prints.Opacity depends on the thickness, refractive index of pigments and brightness of the paper. Brightness of paper in turn can be increased using whitening agents. In the laboratories of paper mills routine opacity measurements are taken from the products. ISO standard of opacity of paper is based on incident diffuse light which is recommended to be detected at zero angle $(D/0°)$. ISO opacity is defined as the reflectance ratio

$$OP = \frac{R_o}{R_\infty},$$
(2.30)

where R_o is the reflectance detected from a single paper sheet (background black) and R_∞ is the reflectance detected from a stack of paper sheets. Human vision, i.e., the response curve of the eye, is important for the appearance of opacity of paper.

A model devised by Kubelka and Munk (K–M) [27] has been used for the description of opacity of paper and other porous media. Detailed description of the K–M model and derivation of their formula can be found from [28]. In the K–M formula the ratio of absorption and scattering coefficients, K/S, is coupled with R_∞ as follows:

$$\frac{K}{S} = \frac{(1 - R_\infty)^2}{2R_\infty}.$$
(2.31)

Equation (2.31) is the basis, e.g., in optical measurement of moisture of paper. Moisture detection is based on the use of water peaks in the infrared spectral region. In the event of fluorescent porous objects, care has to be taken in the interpretation of diffuse spectra. In Fig. 2.11 we show diffuse reflection spectra gained from two different papers obtained from paper mill. It is evident from Fig. 2.10 that different paper grades cause different spectral features in paper. Such data is valuable, e.g., in research and development of paper products.

Figure 2.12 presents a modern multimode laboratory device for paper mills. There are various mechanical and optical measurement modes in the sensor head of Fig. 2.12, among others, the opacity of the paper.

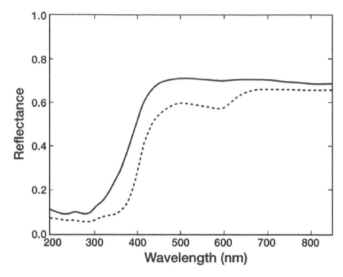

Fig. 2.11. Diffuse reflection from two different paper grades

Fig. 2.12. Measurement head of PaperLab multifunction device for paper quality
(Photo courtesy of MetsoAutomation)

2.5 On Estimation of Optical Constants of Porous Media

The overwhelming problem with porous media such as paper is usually the
multiple scattering of light. Simple models for light scattering are usually
insufficient for rigorous description of the interaction of light with porous

media. Yet, there is a desire to gain information also on the optical constants of such media. In the event that the pores or inclusions in a host media are so small that light scattering is negligible, one may try to utilize effective medium theory for the assessment of the optical constants, such as the effective complex refractive index, of the medium. Information on the effective complex refractive index can be obtained via measurement of reflectance and transmittance. A classical model that is often utilized is based on the effective medium theory devised by Bruggeman [29]. In this model one has to know a priori the volume fraction of the pores in the matrix; the diameter of the pore is disregarded. The model is suited best for the analysis of optical data of nanostructured media. In the case of spherical pores in two-component medium, the effective complex permittivity of the Bruggeman media is as follows:

$$f_h \frac{\varepsilon_h(\omega) - \varepsilon_{\text{eff}}(\omega)}{\varepsilon_h(\omega) + 2\varepsilon_{\text{eff}}(\omega)} + f_i \frac{\varepsilon_i(\omega) - \varepsilon_{\text{eff}}(\omega)}{\varepsilon_i(\omega) + 2\varepsilon_{\text{eff}}(\omega)} = 0, \tag{2.32}$$

where ε_h and ε_i are the complex permittivities of the two components, respectively. The volume fractions of the two components fulfil the relation $f_h + f_i = 1$. If the shape of the pore is different from a sphere, (2.32) can be generalized as follows [30]:

$$f_h \frac{\varepsilon_h - \varepsilon_{\text{eff}}}{\varepsilon_{\text{eff}} + g(\varepsilon_h - \varepsilon_{\text{eff}})} + f_i \frac{\varepsilon_i - \varepsilon_{\text{eff}}}{\varepsilon_{\text{eff}} + g(\varepsilon_i - \varepsilon_{\text{eff}})} = 0. \tag{2.33}$$

where factor g is a shape factor, which is equal to 1/3 for spherical pores. Effective permittivity can be solved from (2.33) as

$$\varepsilon_{\text{eff}} = \frac{-h + \sqrt{h^2 + 4g(1 - g)\varepsilon_i\varepsilon_h}}{4(1 - g)}, \tag{2.34}$$

where

$$h = (g - f_i)\varepsilon_i + (g - f_h)\varepsilon_h. \tag{2.35}$$

The concept can be generalized to hold for multiphase systems and anisotropic media. Unfortunately the effective medium model of Bruggeman is not valid for a great variety of porous products where the simple, different effective medium models cannot be applied. However, there is another approximate model that makes it possible to assess effective optical constants of light scattering media. It is based on the Wiener bounds [9] for a multiphase system which is as follows:

$$\frac{1}{\sum\limits_{j=1}^{J} \frac{f_j}{\varepsilon_j}} \leq \varepsilon_{\text{eff}},$$

$$\varepsilon_{\text{eff}} \leq \sum_{j=1} f_j \varepsilon_j, \tag{2.36}$$

where the permittivity ε_j may be a complex number. Since fill fractions fulfill the condition $\sum_j f_j = 1$, the f_js forms a barycentric coordinate system, which is useful in the estimation of effective permittivity of multicomponent composites [31].

The point of the Wiener bounds is that essentially the medium is considered as two extreme cases namely capacitors having parallel dielectrics or dielectrics in series. Such a treatment may yield rather tight upper and lower bounds for the wave-length dependent effective refractive index of media, and the inclusion itself. This was demonstrated in assessing the refractive index of concentric spherical shell of a copolymer, which is a plastic pigment for paint- and paper industry [32], and for assessment of optical properties of nanoparticles for nanomedicine [33].

2.6 Nonlinear Optical Spectroscopy

In the case of linear optical spectroscopy, the intensity of the probe light is so weak that optical properties of the medium do not change as a function of the amplitude of the light field. However, if we use high intensity lasers as light sources then there may appear observable dependence of the optical properties of the medium on light intensity. Thus, we may utilize such lasers for the study of nonlinear optical properties of media. This possibility opens a relatively wide window to gain various kinds of information on the medium to be inspected because the number of different kinds of nonlinear processes is rather high. We recommend readers interested in nonlinear optical processes to consult the text books of Shen [34] and Boyd [35]. Tunable lasers, where both intensity and wavelength of the radiation can be tuned, offer us means to gain nonlinear optical spectra. Nowadays, femtosecond lasers are popular for experiments in the field of time-resolved spectroscopy. Unfortunately, in most cases the lasers are expensive and experimental setups complicated to measure nonlinear optical spectra. Therefore, it may take time before these devices become popular for routine material analysis in industrial environments. Nevertheless, there is continuous progress towards valuable diagnostic devices in the field of life sciences. Microscopy, which makes use of nonlinear light interaction with biological samples, has been developed. Here we deal only with the case of fluorescence that is generated by two-photon absorption.

The interaction of weak or strong electric field with media can be described with the aid of the susceptibility (χ) of the medium. This susceptibility, which is related to the microscopic properties of the medium, comes from two contributions namely the linear susceptibility (χ_L) and nonlinear susceptibility (χ_{NL}). The former is always present, and it is closely connected to the complex refractive index of the medium. Information on the latter can be obtained in the presence of strong electric field only. According to the nonlinear processes, the nonlinear susceptibility has sub-divisions such as second-order $(\chi^{(2)})$, third-order$(\chi^{(3)})$, ..., multiple-order susceptibility $(\chi^{(n)})$. The strength of the interaction is described with the aid of polarization (P) of electric charges, typically electrons, as follows:

$$P = \chi^{(1)}E + \chi^{(2)}E^2 + \chi^{(3)}E^3 + \dots, \tag{2.37}$$

where E is the amplitude of the electric field. Second-order processes appear only for media where the inversion symmetry of the potential function of the electrons is broken. An important application of the second-order susceptibility is the surface analysis on an interface between two different media. Third-order processes appear with all materials, i.e., those having either isotropic or anisotropic structure.

Two-photon absorption is governed by the third-order nonlinear susceptibility of the medium and in this case the Beer–Lambert absorption law has to be modified as follows [9]:

$$I = \frac{\frac{\mu}{\gamma}}{\left(1 + \frac{\mu}{\gamma I_o}\right)e^{\mu d} - 1}, \tag{2.38}$$

where μ is the linear absorption coefficient, and γ is the absorption coefficient of two-photon absorption, which depends on the wavelength of light. The energy diagram of two-photon excitation and emission is illustrated in Fig. 2.13. The point is the simultaneous absorption of two photons at the same place.

By focusing laser radiation, obtained from a solid state laser, one can choose the volume inside the medium where the nonlinear two-photon absorption occurs. This technique has the advantage that the absorption of the laser line is negligible. In a chromophore two-photon, excitation may induce fluorescence. Thus it is possible to perform, for instance, bioaffinity assaying

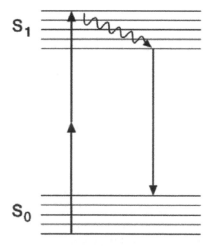

Fig. 2.13. Two-photon excitation and emission

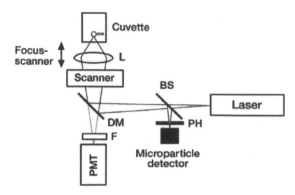

Fig. 2.14. Bioaffinity assaying using two-photon fluorescence. BS, beam splitter; PH, pinhole; D, dichroic mirror; F, filter; and PMT, photomultiplier

with the aid of the two-photon excitation [36]. Such a device is believed to have applications in drug discovery and clinical tests at hospital laboratories. One advantage of this apparatus is that time-consuming preparation of blood samples is not necessary. In Fig. 2.14 we show a schematic diagram of the two-photon fluorescence apparatus [36], which can be used in cytometric studies for detection of particles in turbid matrix.

Before closing this section we remark that conventional fluorescence, i.e., light emission from chromophores at wavelength that is higher than the excitation wavelength, has been utilized both in engineering and life science for the detection of organic media. Commercial devices based on fluorescence have been developing strongly in the field of drug discovery. The key word in drugs discovery is the high-throughput.

2.7 Conclusions

Optical spectroscopy is a well-established technique. Nevertheless, there is a strong development of small size and inexpensive spectrometers for materials inspection both in industry and life sciences. There will be great challenges ahead with optical spectroscopy, especially in monitoring of pollution of the environment caused by industry and the society. We believe that optical spectroscopy will provide at least a partial solution for monitoring the condition of the environment. One can easily understand that due to climate change, the quality of air, drinking water and water bodies such as lakes, rivers and coastal waters is already a big issue. Sensors based on optical spectroscopy are already on markets for on-line monitoring of water quality. A future trend will be that robust optical sensors, using spectra analysis, can be located in remote places, while on-line data and power consumption of the sensor is realized either using a solar cell or wind power or both. These sensors provide real time wireless information on water quality using technology related to cellular phones. Early warnings reach the cellular phone of relevant persons, and in case of emergency a text message to these phones can reach all persons who may suffer contaminated water.

In future the development of small size spectrometers kind of micro machines which can be send, for example with industrial process water for sensing purpose, and which sensor may vanish once the duty cycle is over, will improve real time process monitoring. In the case of a closed water cycle, which would reduce the current use of huge amount of water in some industrial processes, role of reliable sensors for water quality inspection will be very crucial.

3

Machine Vision Systems

Thanks to the fast development of robust CCD- cameras the field of machine vision technique for process and product monitoring has become very popular in different sectors of industry. Typical applications of machine vision include tasks such as inspection, measurement, recognition and positioning of products among the others. It would be a topic of another book to describe the wide field of machine vision. Here we focus only on few applications. Some descriptions of systems that practically speaking utilize machine vision appear also in other chapters of this book.

3.1 Inspection of Plastic Cover of Mobile Telephone

The machine vision system usually involves a passive or active light source for the illumination of the object, one or more CCD-cameras, optical elements and a frame grabber, which itself can process the image or it can transfer information of the image into the central unit of a personal computer (PC). In some applications the CCD-camera may be installed into the arm of an industrial robot to check, e.g., the quality of laser welding in the automotive industry. An important requirement of a machine vision system is usually to arrange appropriate illumination of the object by a controlled light source so that a sharp image of the object can be obtained. Different external disturbances, for example, low light intensity, fluctuation of the light intensity and stray light may have an effect on the detection of the object. In choosing the geometry of illumination one has to take care of things like the contrast of the image, shadows, edges and also the movement of the object.

In Fig. 3.1 is shown one example of a schematic diagram of a machine vision system where a plastic cover of a mobile phone is inspected. A high quality of the cover is required, therefore, in the case of Fig. 3.1 an inspection of the cover may include tasks, such as, the condition of the protection tape attached on the display, lens of a mobile camera, telephone and dust seal, respectively. One may expect further development in the field of mobile telephones, and

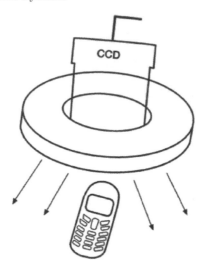

Fig. 3.1. Optical inspection of plastic cover of mobile phone. The light source takes the form of a ring and CCD camera is used as a detector for image analysis

therefore their quality inspection by optical and other means becomes more popular. It is possible to implement optical and other sensors into the mobile telephone, and these sensors may in future have different kinds of functions, e.g., in increasing the security of people.

3.2 Diffractive Optical Element Based Machine Vision Gauge for Float Glass Thickness Measurement

Diffraction can be considered as one of the basic optical phenomena. It usually presents regularity in the light pattern observed at the far-field region from an obstacle that produces diffracted light wave intensity pattern. Hence, diffraction is usually easier to characterize than scattering of the light. Diffraction appears both for incoherent, partially coherent and coherent radiation, and it appears when light is transmitted or reflected from micro structures. In engineering applications diffraction is quite often exploited in the spectral region of visible light.

3.2.1 Diffractive Optical Element

Diffractive optical elements (DOE) have been known for a relatively long time. A grating is perhaps the simplest example of the DOE, and grating as an alternative of a prism has found applications especially in optical spectroscopy. Indeed with a grating one can disperse white light into its spectrum. Nowadays, holographic gratings are widely used in spectrophotometers because of their good dispersion resolution in the monochromator. The measurement time of a spectrum can be made much shorter with a grating than a prism spectrometer. In the case of prism one has to rotate the prism, whereas with a grating the whole spectrum is obtained immediately with a modern array of detectors.

In a general case the DOE may change both the amplitude and phase of the incident light. The light modulation by diffraction can be due to the local change of the surface topography and the refractive index of the DOE.

3.2.2 Float Glass

Sheet glass is manufactured by floating glass ribbon on molten purified tin, thereafter it is drawn, cooled and cut at the process line. Various factors affect the quality of the float glass, and each factory has its own recipe. Thus, for instance by optical spectroscopy it is possible to find differences between products from different sources. Float glass is often subject to further processing such as tempering, toughening and lamination in order to increase its strength and safety properties. Float glass has many applications in addition to the windows of buildings. The highest quality float glass is needed in scientific glasses and car wind shields. The thickness of the electronic glass is usually less than one millimeter, but in most of the applications several millimeters.

There exist different kinds of optical off- and on-line gauges at the glass factories for the evaluation of the quality of float glass. Probably the most important quality parameter is the thickness of the float glass. For that purpose a sensor based on laser beam reflection from upper and lower surfaces of the glass ribbon has been utilized [37]. Due to the demand on high accuracy of float glass thickness measurement, another type of sensor was constructed for on-line operation [38]. This sensor is based on the exploitation of diffractive optical element (DOE). In addition to the thickness signal the DOE sensor gives information on the edge distortion of the ribbon. In the machine direction the ribbon moves forward on a roller conveyor. The rotators leave traces on both edges of the solidifying glass. Thus edge distortion, which takes a wedge shape, appears. This edge distortion can be measured by observing optical power, which is given by a millidiopter (mdpt) for the float glass. The edge distortion is important in the estimation of the cutting line of the ribbon at the both edges. If the edge distortion were rather high then, e.g., in a wind shield of a vehicle the driver may observe an object distorted or apparently moving although the object is at rest. Therefore, it is easy to understand that high quality float glass is of crucial importance, e.g., in improving security of vehicles.

3.2.3 DOE as an On-Line Thickness Gauge of Float Glass

The DOE in the present case is a computer-generated hologram [39]. The DOE, presented in Fig. 3.2, was calculated by using Rayleigh–Sommerfeld diffraction integral [40], and it is an on-axis binary-amplitude hologram whose

Fig. 3.2. Diffractive optical element

transmission is 50%. The output pattern of the planar DOE, when constructed by an expanded laser beam, is a regular array 4 × 4 light spot matrix. The size of the aperture and the focal length of the DOE, as well as the output pattern can be designed according to the application.

The DOE in Fig. 3.2 is sensitive both on the amplitude and the phase of the reconstructing wave. This element was produced using electron beam lithography. A chrome layer was sputtered on glass substrate, and a positive electron resist was deposited on the chrome layer, which in turn was exposed to the e-beam writer. A chrome mask was obtained after development of the resist and wet-etching the chrome. The prize of the DOE is high because e-beam writer and other facilities needed are expensive, however, the prize can be reduced if the master element is copied in mass production. In the present case the aperture size of the DOE is $8 \times 8\,\text{mm}^2$ and the focal length is 200 mm. The size of individual light spot is about 30 μm. The advantage of the small spot size is a better accuracy in the thickness measurement, and a high number of light spots provide means for reliable statistical analysis using the image data obtained by the gauge. Note that each individual light spot can be used in the calculation of the glass thickness by using the setup shown in Fig. 3.3. The expanded laser beam is first incident on the DOE.

The diffracted wave front is reflected from the upper and lower surfaces of the float glass. The two spot matrix patterns are recorded simultaneously at different locations of the cell of the CCD camera, i.e., there is no objective lens at all. Due to the low speed of high quality float glass production line, real time measurement of glass thickness is possible.

The glass thickness (d) is obtained using the notations of Fig. 3.3 and Snells's law, as well as trigonometry. Snell's law can be expressed as follows:

$$\beta = \arcsin(\sin\ \alpha/n), \qquad (3.1)$$

where n is the refractive index of the glass. From Fig. 3.3 we observe that

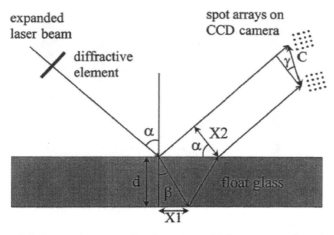

Fig. 3.3. Schematic diagram of a float glass thickness gauge based on DOE

$$x_1 = d \tan \beta$$
$$x_2 = 2x_1 \cos \alpha. \tag{3.2}$$

Now it is important to realize that the CCD camera in Fig. 3.3 has to be tilted so that both reflected light spot patterns would be in focus. Thus the distance on the chip between the two spot matrix images reflected from the upper and lower surfaces of the float glass is equal to

$$C = \frac{x_2}{\cos \gamma}. \tag{3.3}$$

Next we get from (3.1) to (3.3)

$$C = \frac{2 \tan [\arcsin(\sin \alpha/n)] \cos \alpha}{\cos \gamma} d. \tag{3.4}$$

The factor C of the glass thickness in (3.4) depends both on the refractive index of the glass, which for high quality glass is practically speaking a constant, and the measurement geometry. This factor can be optimized so that minimum error is obtained for the thickness. The accuracy of thickness measurement with the DOE gauge is $\pm 10\,\mu\text{m}$.

In industrial sites the sensor head is scanned over the glass ribbon. The temperature in the float glass production line is rather high therefore, cooling of the sensor head is important, and it can be cooled by circulating water, or using air cooling. In Fig. 3.4 is shown, as an example, grey scale (it can be given colour-coded) map of float glass thickness for a 2 km long ribbon.

A future trend regarding the glass industry might be an optical and multi measurement on-line gauge that provides simultaneously any information on the defects, thickness, optical power, surface roughness, spatial refractive index variation and other relevant properties of the glass ribbon. Such a measurement technique has to be based on probing instantly a large area.

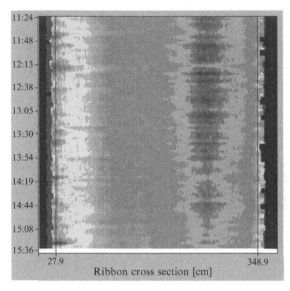

Fig. 3.4. Grey scale thickness map of float glass

3.3 Machine Vision System for Monitoring Compressed Paper

The number of paper products is largely due to different applications. Paper makers pay continuous attention to improve the quality of paper, and pigment producers play an important role in this process. For the measurement of the smoothness of the paper there is the Chapman's optical method and device [41], which is well-known in paper science. It makes use of the total reflection of light in it's assessment of the contact area between the paper and a probe prism. In a way this measurement technique simulates closely the situation between paper and a nip of a paper machine. Since many paper grades, such as, Xerox-, news-, fine- and super-calendered (SC) paper, are subject to be printed, optical measurement techniques for simulation of printing process have been introduced, using, e.g., laser beam reflection [42]. Here we describe a rather simple machine vision technique, along with a compression system of the paper. The principle of the device is illustrated in Fig. 3.5.

In Fig. 3.6 we show examples of image data obtained for fine paper at the same locations when the compression pressure is increased. From Fig. 3.6 we observe that the area of dark spots has increased as a result of increasing the compression that the paper experiences between the two windows. The dark patterns are due to the contact of the paper with the upper probe window, a phenomenon which was confirmed by a spectroscopic study [43]. The image information, such as in Fig. 3.6 is important in the research and development of the surface roughness, formation and compressibility of different paper grades.

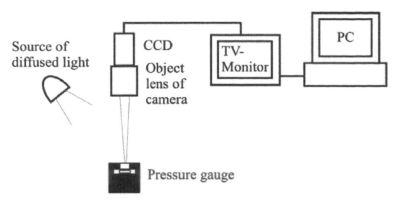

Fig. 3.5. A schematic diagram of machine vision system for surface inspection of compressed paper

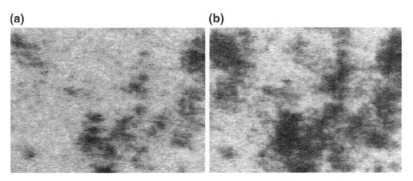

Fig. 3.6. Dark areas on compressed paper surface (**a**) pressure equal to 1.2 MPa, and (**b**) 2.4 MPa. The area of the image is $2.85 \times 3.8\,\text{mm}^2$

Paper industry is one of the most advanced branches as it concerns sensing of the process and product during the paper making process. In future also in this industrial sector, demands on instant multi measurement gauges will increase. Simultaneous measurement of moisture, surface roughness, absolute gloss, gloss mottling, opacity and other relevant paper properties from a large area is a great challenge, especially because the machine speed of modern paper machines is very high ca. $2\,\text{km}\ \text{min}^{-1}$.

3.4 Imaging Spectrometer

With the aid of a CCD-camera and a spectrometer it is possible to construct a devise that gives information both on image and spectra from various kinds of objects ranging from UV- up to infrared radiation. By the imaging spectrometer one can gain a two-dimensional image from the object as well as a spectrum from the image pixel. These devices image either two spatial dimensions

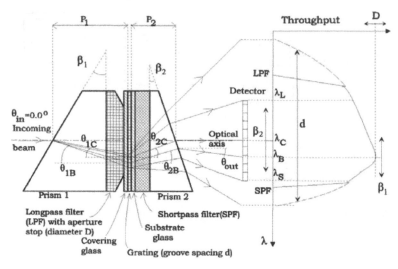

Fig. 3.7. Schematic diagram of PGP imaging spectrometer (by the courtesy of Dr. Mauri Aikio)

and temporally sample the spectrum, or image one spatial dimension and the spectrum while sampling the other spatial dimension.

As it concerns applications airborne systems that perform remote sensing have been realized. In addition imaging spectrometers can be utilized for inspection tasks in the industry such as controlling process and product quality, and also in medical imaging.

For example, one can record reflection spectrum from an object and calculate the colour coordinates for each pixel of the image.

Here we present an imaging spectrometer that is based on the prism-grating-prism (PGP) construction [44]. The optical implementation of the PGP imaging spectrometer is shown in Fig. 3.7. It samples one spatial dimension and the spectral dimension simultaneously. The second spatial dimension is obtained by scanning the field of view.

The principle of the operation is that each element in the image provides its intrinsic spectrum. The volume transmission grating between the prisms may be, e.g., dichromated gelatin. The role of the diffraction grating is to sort spatially the different wavelengths. The spatial spectrum is analyzed. Naturally, the outcome is usually a huge data set. In addition to the conventional image one can display a two-dimensional Cartesian co-ordinate system where one axis is the spatial axis and the other is the spectral axis. In Fig. 3.8 is shown a two-dimensional RGB-image, which was calculated from the spectral image, of a print on a paper with blue ink density mottling. A reflection spectrum from one location on the paper is also shown in Fig. 3.8. The pixel size was $117 \times 117 \, \mu m^2$, and the spectrum was obtained at the visible range with 5 nm step.

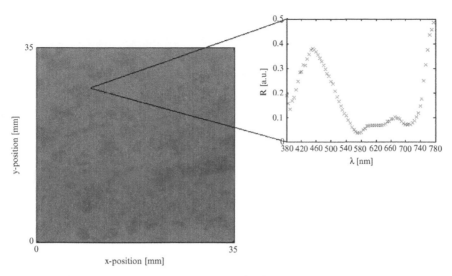

Fig. 3.8. A print with density mottling (*left*) and a reflection spectrum from one pixel

3.5 Conclusions

Imaging technique will become more popular in an industry that has already adopted machine vision, e.g., in inspection of mobile phones. Imaging and sensing will be a future challenge in nanomedicine because information on small features is required. A trend will be that imaging system has to be smaller and faster than it is nowadays, and they have to provide real time multiple-information. Key words in the industry are reliability, cost effective and on-line operation of a measurement device in harsh environments.

4

Optical Inspection of Surface Roughness and Gloss

For a variety of products, presenting different industrial sectors, the quality of the product surface is quite often an important factor. The product surface may go through a sequence of finishing processes in order to improve the surface quality. In this chapter we mention a few examples where surface quality is a critical parameter. For instance, in the case of steel sheet production for the automotive industry a relatively thick steel strip is first hot-rolled and thereafter cold-rolled in a roller line. The rollers leave characteristic finishing marks on the surface of the steel strip. The surface profile of the roller surface is essential, in addition to the rolling load, as it concerns the desired surface roughness and surface texture of the steel sheet. Naturally there appears a wear on the envelope surface of the roller, and this wear causes the surface roughness of the metal strip to change as a function of the wear of the roller. A similar situation may appear also in another type of surface finishing, namely in the production of tablets in the pharmaceutical industry. There the surface finish of the punches of a tablet machine is crucial in order to prevent, e.g., sticking of the powder material onto the surface of the punch. Thus a very smooth punch surface is usually required for optimum tablet making. In the case of cellular phones high quality plastic covers are obtained using electro-discharge machining for injection mold surfaces. Plastic melt copies the mold surface and if wanted a relatively rough surface matte can be obtained. However, the tool is subject to wear and may cause local surface quality variation in the plastic product during a long duty cycle of the tool.

It is obvious that in some sectors of the industry there is a desire to get as smooth a product surface as possible, whereas in some other sectors a relatively high surface roughness is the measure of the quality of the product. A common factor for all such industrial branches is to get information of the surface roughness of the product, and also on the finishing marks and their orientations. The role of the manufacturing process and, for instance, the condition of the tool has great importance in finding the optimal process in order to gain optimal product surface condition [45]. Optical measurement techniques for ultra smooth surfaces is of great importance especially in the optimization of the surface quality of devices based on semiconductors and other high tech products. Optical measurement of ultra smooth surfaces have been described, e.g., in the book of Bennett and Matson [46].

4.1 Definition of Surface Roughness Parameters

Measurement of surface roughness parameters, which we will define below, has been traditionally based on a device called diamond stylus. The idea is to apply a small force on a thin diamond stylus so that it records surface profile of the surface along a thin, and usually rather short line. Unfortunately, such a technique is restricted to laboratory conditions, and it cannot be considered for on-line inspection of the product's surface quality at the industrial site, where the line speed can be rather fast. Fortunately there are solutions for nondestructive surface roughness estimation that are based on optical measurement techniques such as triangulation, specular reflection and also laser stylus. Such measurement techniques provide macroscopic information on the surface roughness which certainly is appreciated among quality inspectors in the various sectors of the process industry.

In engineering the most popular surface roughness parameters are the average surface roughness (R_a) and root-mean-square (rms) surface roughness (R_q). They are based on purely mathematical definitions as follows:

$$R_a = \frac{1}{A} \iint_A |z(x,y) - \langle z(x,y)\rangle| dx\ dy, \tag{4.1}$$

and

$$R_q = \left\{ \frac{1}{A} \iint_A [z(x,y) - \langle z(x,y)\rangle]^2 dx\ dy \right\}^{1/2}, \tag{4.2}$$

where $z = z(x,y)$ is the height as a function of location in Cartesian coordinate system, and $\langle z(x,y)\rangle$ is the mean surface in the xyz- space chosen in a manner that $z(x,y)$ has a minimum variance. The scanned area is A. Note that in practical measurements the scanning of a surface topography is fixed so that the lag length along a line, say in the x-direction, is a constant and the scanning of the area A is based on a step wise line scanning procedure. Quite often the one-dimensional counterparts of the definitions (4.1) and (4.2) are utilized. That is to say the profile is $z = z(x)$ because of the measurement of the surface profile along a thin line only. Thus the integrals needed for the calculation of the surface roughness parameters involve only one variable. Unfortunately, the definitions of (4.1) and (4.2) are not free of problems. For instance, say we rotate 180° a measured one-dimensional profile $z = z(x)$ with respect to the x-axis. Due to the definitions (4.1) and (4.2) we get the same values for the surface roughness parameters both for the original and the rotated mirror profile. However, their mechanical functions

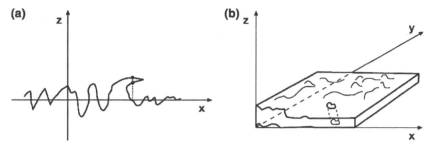

Fig. 4.1. Pathological cases for definition of surface roughness. (a) Definition of surface roughness is unambiguous in the location of the *dotted line*, and (b) in the location of a "worm hole"

may be rather different, for example, in lubrication applications. Thus the standardized surface roughness parameters and practical engineering do not always meet each other. There are also other problems which are illustrated in Fig. 4.1. In Fig. 4.1a we have a case where the profile function z has two different values at the same point. Such a situation is intolerable in the sense of mathematical definition of a function z, but such finishing marks may appear in practice. Also in the surface profile measurement, either by mechanical or optical stylus, the space under the "horn" in Fig. 4.1a is difficult to access. Another problematic situation is shown in Fig. 4.1b where a "worm hole" appears. Such a situation may appear in the context of a porous media such as the paper. In this case the definition of the height is questionable inside the area of the "worm hole". Despite some problems in the definition of the surface roughness parameters they provide good means for quality inspectors to assess especially planar products at the industrial site. However, measurement of the surface roughness of complex shaped objects, for example, the interior wall of a car engine cylinder remains still as a problem.

In a similar manner as above we can also define slope parameters related to the surface roughness using the concept of partial derivatives as follows:

$$S_{a,x} = \frac{1}{A} \iint\limits_A \left| \frac{\partial z(x,y)}{\partial x} - \frac{\partial \langle z(x,y) \rangle}{\partial x} \right| dx\, dy$$

$$S_{a,y} = \frac{1}{A} \iint\limits_A \left| \frac{\partial z(x,y)}{\partial y} - \frac{\partial \langle z(x,y) \rangle}{\partial y} \right| dx\, dy \tag{4.3}$$

and

$$S_{q,x} = \left\{ \frac{1}{A} \iint\limits_A \left[\frac{\partial z(x,y)}{\partial x} - \frac{\partial \langle z(x,y) \rangle}{\partial x} \right]^2 dx\, dy \right\}^{1/2}$$

$$S_{q,y} = \left\{ \frac{1}{A} \iint\limits_A \left[\frac{\partial z(x,y)}{\partial y} - \frac{\partial \langle z(x,y) \rangle}{\partial y} \right]^2 dx\, dy \right\}^{1/2} . \tag{4.4}$$

Due to the experimental restrictions and also partly due to the practical easiness the slope parameters are usually obtained only in one dimension while in the other direction it is a parameter. In the general case the slope should be treated using the directional derivative, i.e., the gradient. Same problems as above hold in the definition of the slope parameters for "badly behaved". The slope parameter is a practical measure in the sense that the two surfaces may have the same average surface roughness but they may differ from each other because of different average slopes.

Other useful parameters, which however may be less exploited in the daily engineering practice, are the autocorrelation (AC) and power spectral density function (PDF). The autocorrelation function quantifies the similarities of the surface profile in lateral direction. PDF is obtained by the squared modulus of a Fourier transform and describes especially periodicity of finishing marks in the spatial frequency plane (f_1, f_2). These functions, which are of similar nature, are defined in the general case by (4.5) and (4.6)

$$\mathrm{AC}(\tau_1, \tau_2) = \frac{1}{A} \iint\limits_A z(x,y)z(x + \tau_1, y + \tau_2)\mathrm{d}x\ \mathrm{d}y, \qquad (4.5)$$

$$\mathrm{PDF}(f_1, f_2) = \frac{1}{A} \left| \iint\limits_A z(x,y)\mathrm{e}^{2\pi i(xf_1 + yf_2)}\mathrm{d}x\ \mathrm{d}y \right|^2 . \qquad (4.6)$$

Typically AC and PDF are calculated for the one-dimensional case. Thus autocorrelation length $l = \mathrm{AC}(0)/10$ is defined as a measure depending on how much the one-dimensional surface profile resembles itself. The surface statistics has an impact on the autocorrelation function [47].

The surface profile is obtained usually by sampling discrete but equi-spaced data points. Thus in data analysis one has to utilize approximations of the above mentioned parameters by replacing the integrals involved with the corresponding sums, or use spline fitting in order to get continuous functions.

Probability theory is useful in the description of the distribution of heights. The distribution function $w = w(z)$

$$\int\limits_{-\infty}^{\infty} w(z)\mathrm{d}z = 1 \qquad (4.7)$$

give information about the symmetry or asymmetry of "hills" and "valleys" that is to say skewness of the distribution. The simplest case is the Gaussian distribution function, i.e., a symmetric bell shaped distribution. Unfortunately, quite often the distribution function of surface heights for various types of product surfaces is asymmetric, for instance, due to finishing process such as flat lapping. It may happen that a simple mathematical expression for the distribution may not be a straightforward issue. As an example of such a situation we remember the case of the complex surface structure of paper. Thus quite often Gaussian distribution can be considered as the first approximation for a probability model of the surface statistics.

4.2 Optical Inspection of Finishing Marks

Surface roughness can be divided roughly into macro- and micro surface roughness. Evidently macro surface roughness is something that is based on visual perception, whereas micro surface roughness requires other means in order to get data on it. Human eye is accurate and has large dynamic range, thus visual inspection has been for a long time a well-established method in the industry for the inspection of a product surface quality. This is true even now, although the demand for automatic inspection, which is based more or less on machine vision has become stronger in many sectors of the process industry. In the case of visual inspection of macro roughness or abnormal curvature of an object an expanded HeNe laser beam provides a nice way to reveal the abnormalities of a surface. Another thing is how to analyze such image data, data that is evident for a human eye, but not for a system based on a CCD-camera and a computer.

For the purpose of monitoring micro surface roughness monochromatic and coherent light radiation from various cheap laser sources, including also semiconductor lasers, provides a probe for the inspection of a variety of different surfaces. One important practical factor is usually the low divergence of the laser beam. This laser beam property enables remote measurement and detection. Also the monochromatic light is important especially in the data analysis which involves relatively simple mathematical models that usually assume use of monochromatic light. In addition to the micro surface roughness measurement a laser beam can be exploited as a light source for the detection of some preferred direction of finishing marks, which may also be detected by visual inspection, e.g., in the case of cold-rolled metal products one can distinguish the finishing marks by a naked eye. The finishing marks constitute usually a nonperfect grating structure. Thus the laser beam is diffracted from such a reflection grating. In Fig. 4.2 is shown, as an example, a blazed groove profile of a reflection grating. Maximum in the diffracted light intensity at the far-field region is obtained if the condition

$$m\lambda = d(x)(\sin \theta(x) \pm \sin \beta(x)), \qquad (4.8)$$

is valid. In (4.8) m is the order of the diffraction, λ is the wavelength of the laser, θ is the angle of incidence and β is the angle of diffraction. The plus sign is applied if the incident and diffracted ray is on the same side of the normal to the grating. Due to the irregularity of the grating, that is to say the finishing marks, all the grating parameters, depend on the location at x-axis, as it is illustrated in the example of Fig. 4.2. Generally speaking, for rigorous mathematical treatment of the diffraction pattern the grating has to be dealt

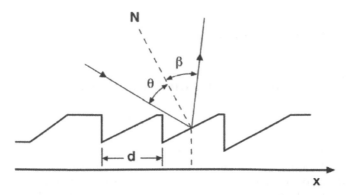

Fig. 4.2. Blazed groove system of a reflection grating

Fig. 4.3. Diffraction pattern from cold-rolled steel. The light source was a HeNe laser

with as a three-dimensional system. Because of the nonperfect reflection grating of a surface of a metal or other products one usually observes a pretty large number of diffraction orders that span quite often a large angle in the far-field region. For instance in the case of parallel flat lapping one observes in the far-field region, provided that the laser beam is applied in perpendicular direction against the groove structure, a line full of different diffraction orders and also speckled pattern around the line. In the event of criss-cross flat lapping the diffraction pattern imitates the finishing process, i.e., two lines covering different diffraction orders that appear in a criss-crossed line image at the far-field region. In the cases described above one may observe the orientation of the finishing marks with a naked eye. For some applications the finishing marks have to be made invisible. Usually this happens by polishing the surface. Nevertheless, even in such a case the original finishing marks can usually be revealed by a simple laser beam scattering experiment. Thus laser light is a very sensitive probe to study the nature of finishing marks and even to reveal hidden marks for visual inspection. It also gives a visual measure about the success of the finishing process. In Fig. 4.3 is shown a diffraction pattern obtained from a cold-rolled steel plate.

The number of simultaneous inspection of different locations of finishing marks can be increased by fanning one laser beam into a punch of several beams by a diffraction grating, or by using a micro lens array. The hard part is the realization of an imaging and analyzing system that should record information from a great number of spots on the object.

4.3 Measurement of Surface Roughness Using Laser Beam as a Probe of Specular Reflection

Some care has to be taken when measuring surface roughness using optical techniques, namely the mechanical surface roughness parameters above, which do not take into account some specific features that appear when using light for surface probing. For instance in the case of relatively rough surface multiple surface- and also in some cases (porous media) bulk scattering of light may have an effect on the signal, and yield an erroneous estimate for the surface roughness. We suggest that the concept of optical surface roughness (R_o) is used instead of surface roughness in some cases, for instance, when detecting a light signal that may carry noise due to single or multiple scattering, i.e., diffuse light. Such a situation appears for instance in the detection of the specularly reflected light from a rough surface. Fortunately, in some cases one can find rather a good correlation between optical and mechanical surface roughness parameters. Thus, in principle two corroborative methods can be utilized, namely optical and a mechanical one (diamond stylus) for the assessment of the surface roughness of nonfragile surfaces. In the design and construction of an optical surface roughness gauge or other optical measurement devises, in order to operate in severe industrial environments, care has to be taken to avoid, e.g., dust particles on the optical elements. This is not always possible. The customer who purchases the optical measurement device is usually instructed how to manage with typical cleaning tasks of the optical elements. Nevertheless, annual or denser inspection of the function of the measurement apparatus may be required by an authorized person.

The specular reflection means that the angle of the incidence and the reflection of the laser beam are the same. In addition, the incident and reflected beams and the normal of the surface are in the same plane. Let us consider next the specular reflection from a rough surface according to Fig. 4.4. A plane wave from the laser is incident on the surface. Due to the surface roughness there will be scattering of light, which means that in addition to the specularly reflected light there will appear diffused light that propagates into a hemisphere. For the sake of simplicity in Fig. 4.4 we have drawn two adjacent light rays. Due to the surface roughness, i.e., height z, there will be optical path difference, which is obtained by trigonometry, between the two rays equal to the expression

$$\Delta s = 2z \sin \theta, \tag{4.9}$$

where θ is the angle of incidence, and it is equal to the angle of reflection. The optical path difference can be restated with the aid of the phase difference. Thus we can write

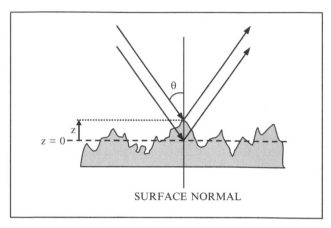

Fig. 4.4. Specular reflection of two adjacent laser light rays from a rough surface

$$\Delta\phi = \frac{2\pi}{\lambda}2z \ \sin \ \theta, \tag{4.10}$$

where λ is the wavelength of the laser. The two adjacent rays interfere and their resultant field after reflection, provided that the rays preserve their plane wave forms, is

$$\overline{E} = \overline{E}_o e^{i(\overline{k}\cdot\overline{r}-\omega t)} + \overline{E}_o e^{i(\overline{k}\cdot\overline{r}-\omega t+\Delta\phi)}, \tag{4.11}$$

where $\overline{E}_o$ is the amplitude of the electric field, $\overline{k}$ is the wave vector, $\overline{r}$ is the position vector, ω is the angular frequency of the monochromatic laser radiation, and t denotes time. From (4.11) we obtain a useful form, typical for interference, namely the intensity of the light (I) which is proportional to

$$I = \left|\overline{E}\right|^2 = \overline{E}\cdot\overline{E}^* = 2\left|\overline{E}_o\right|(1 + \cos \ \Delta\phi). \tag{4.12}$$

From (4.12) we can conclude that maximum intensity is obtained when $\Delta\varphi = 0$, i.e., in the case of perfectly smooth surface, whereas minimum is obtained when $\varphi = \Delta\pi$. This latter condition means a totally rough surface. Unfortunately due to our simple qualitative model, which involves periodic cosine function, interference maximum is obtained for integers of 2π. Of course such a situation is absurd in practice. The question is then how to construct a device so that the working range is always well below the upper limit of the phase angle namely π. There are two possibilities to find such a working range. One has to remember that the surface roughness range of planar objects may be known prior due to the product specifications settled by the manufacturer. This provides frames for the measurement parameters of the device. From (4.10) we can deduce that the phase angle is decreasing either if we let the wavelength of the laser or the angle of the incidence to increase. From the technical point of view the wavelength of cheap lasers is usually fixed so the easier way is to affect to the angle of the incidence.

Next we deal with a popular model of Beckmann and Spizzichino [48] that has been utilized in the surface roughness measurement of metals especially, and Cielo [49] has presented its implementation in the metal industry. The model leans heavily on the Gaussian surface roughness statistics. The assumptions are that we are dealing with an ideal conductor, the surface height distribution and the autocorrelation function obey normal distribution, and the curvature of the surface is much larger than the wavelength of the light. A somewhat lengthy derivation of the result (4.13), for rough surface generated by random processes, can be found from the book of Beckmann and Spizzichino (see also [50])

$$I = I_{\mathrm{o}} e^{-\left(\frac{4\pi R_{\mathrm{o}} \cos\theta}{\lambda}\right)^2}, \tag{4.13}$$

where it is assumed that a perfectly smooth surface reflects 100% of the incident light, and I_{o} is the intensity of incident laser beam. In a general case when the surface roughness probability distribution function is not known we may express the detected light intensity formally by a functional dependence

$$I = f(N, \theta, R_{\mathrm{a}}) I_{\mathrm{o}}, \tag{4.14}$$

where N is the complex refractive index of the bulk medium and

$$|f(N, \theta, R_{\mathrm{a}})| < 1. \tag{4.15}$$

In Fig. 4.5 is shown a schematic diagram of the measurement system. The beam splitter is needed for reference signal so the aging of the laser can be cancelled in the signal analysis by detection of the ratio I/I_{ref}. For the assessment of the optical surface roughness one should arrange synchronous scanning of the angles of incidence and the reflection. Thus a goniometer is required. Such procedure is time consuming and the location of inspection may have passed far away if the measurement system is intended for on-line monitoring. Hence

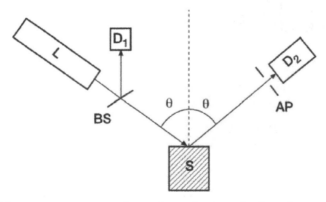

Fig. 4.5. Measurement system of specular reflection of a laser beam. L, laser; D, detector; S, sample; BS, beam splitter; and AP, aperture

the angle scanning mode is feasible only for off-line inspection in controlled laboratory conditions. However, if the requirement of getting accurately the absolute value of the optical surface roughness that is not a critical parameter then one can adjust the angle of incidence and monitor the intensity ratio for assessment of increasing or decreasing of the optical surface roughness. Using one or more lasers working at different angles of incidence it is possible to find, e.g., detection limits for a desired surface roughness of a planar product. Note that the laser spot on the surface is relatively large near grazing incidence, thus the sensor is integrating surface roughness from such an area. Therefore we get an estimate for the average surface roughness. The detector of the measurement system can be a photo diode together with an aperture to block diffuse reflection, or a CCD-camera. The latter one makes it possible to record angular reflection pattern. Thus simultaneous information on the specular and diffuse reflection in the specified geometry is available.

In Fig. 4.6 we show an example of the development of the specular reflection from starch acetate compact as a function of the angle of incidence and using CCD camera as a detector.

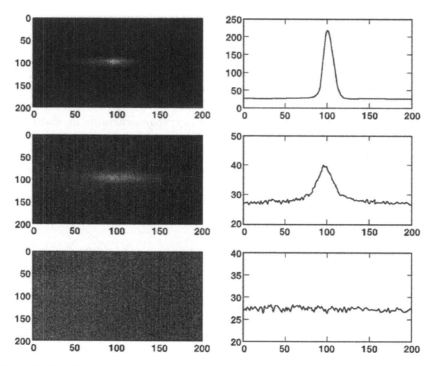

Fig. 4.6. Development of specular reflection pattern from starch acetate compact. Image of reflection pattern (*left*), and corresponding light intensity distribution (*right*) along a line [51]

From the images of Fig. 4.6 we observe that for a relatively low angle of incidence a speckled pattern appears, whereas for a relatively high angle of incidence specular reflection becomes relatively strong.

Quite often a medicine takes the form of a tablet and the porosity and surface roughness of the tablet affect the rate of the release of the drug substance, which is in a matrix such as starch acetate. In the treatment of some diseases it is of critical importance to know this rate of release and absorption of drug substance in order to provide the correct dosing. Tablet coating is an additional procedure and it can be used for various practical factors, such as, to control the release of the drug from the tablet or to protect the drug from the gastric environment of the stomach. Using optical measurement techniques it is possible to evaluate the surface roughness of a tablet with or without coating. The aim of such an inspection is to help tablet makers to optimize the quality of the tablet.

Although the device stated above is based on the use of monochromatic light the spectral properties of metal or other media have to be known, especially if one wants to measure the surface roughness of different materials. This is important in calibration of the system for measurement of the absolute surface roughness.

In Fig. 4.7 we show an example of goniometric data for two flat steel punches, namely a used and an unused one. Such punches are used in the pharmaceutical industry for tablet manufacturing. The logarithm of the intensity ratio is presented as a function of $\cos^2 \theta$. The upper curve indicates that the surface roughness of the unused punch is lower than that of the used one.

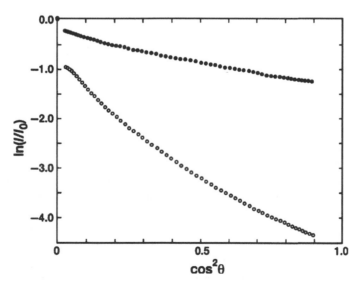

Fig. 4.7. Signal for specular reflection of laser beam from used (*lower curve*) and new (*upper curve*) flat punches used in the pharmaceutical industry [52]

Usually the used punch is rejected or it undertakes a renewal of its surface finish after a duty period. One symptom of the decrease of the quality of the punch is the sticking of the tablet powder on its compression face. It is possible to clean the punch surface by mechanical polishing, or using laser ablation. Note that in the goniometric measurement one indication of the surface smoothness is the level of the signal as that in the Fig. 4.7, and also how a small angle of incidence will still yield a signal. For rough surfaces small angles of incidence yield usually a negligible signal that is mainly noise. Measurement data (Fig. 4.7) can be used also for estimation of the statistics of the surface roughness. Due to the nearly linear data obtained from the logarithm of the intensity ratio of unused punch in Fig. 4.7 we can conclude that the surface statistics for this surface is approximately Gaussian, whereas the corresponding statistics of the used punch has evidently changed due to its wear, and departs from the Gaussian surface roughness statistics.

We mention that DOE has been also utilized for surface roughness inspection both as a fringe projection unit [53] and as an analyzer of scattered light wave, e.g., from metals, paper, tablets, etc. [54].

In the pharmaceutical industry the future trend is the process analytical technology (PAT), an issue that has been for a long time realized in some other sectors of the industry. Important issues in the pharmaceutical industry are noncontact, safe and real time monitoring of the condition of raw materials, success of the process and the final product. Optical measurement technique involving use of lasers as light sources, CCD cameras for image formation and optical-, Raman- and terahertz spectroscopy will satisfy most of the measurement demands, but much work has to be carried out before at least a partial solution for successful PAT has been realized.

4.4 Measurement of Surface Roughness Using Focused Laser Beam

There are various realizations of optical devices for the surface profile measurement using a focused laser beam as a probe. Once the surface profile is measured, various statistical surface roughness parameters can be computed from the measured data. At the industrial site such statistical data provide important information about the history of the quality of the product, and may assist in optimizing the process parameters if a reliable feed back system has been installed. Such a procedure is practical especially when a proper on-line inspection of the quality of the product has been organized.

Here we consider the measurement of surface profile using laser profilometry by confocal measurement and laser triangulation. Both of these techniques involve usually scanning of a thin line and inspection of the laser spot image on the product surface. The lateral resolution is defined by the waist of the laser beam typically between 1 and 10 µm. The feature size on the surface must be larger than the beam waist. In Fig. 4.8 is presented a schematic diagram of

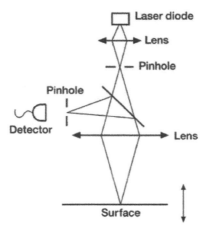

Fig. 4.8. A principle of a laser profilometer

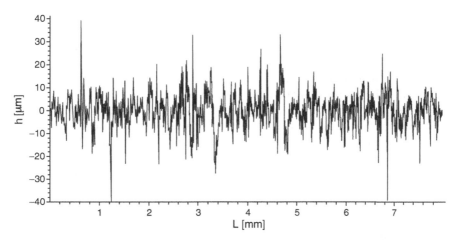

Fig. 4.9. Surface profile from steel sample recorded by a laser stylus

the principle of confocal measurement of the surface profile. The laser beam is incident along the normal of the surface. The light source is a stabilized laser diode and the output beam is focused using a lens system as shown in Fig. 4.8. The beam focus is scanned in vertical direction. Maximum light intensity in reflection is obtained once the "hill" or "valley" coincides with the focal point of the lower lens. In the case of the out-of-focus light intensity that is incident on the photo detector is drastically reduced because of the blocking pinhole. For fast measurement one has to require high measurement frequency and the scanning speed in the vertical direction. In Fig. 4.9 is shown a typical surface profile obtained from a steel product by laser profilometer.

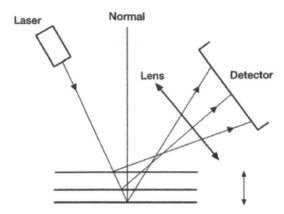

Fig. 4.10. Principle of triangulation. The light beam direction after reflection depends on the surface undulation (here described as a straight line)

In Fig. 4.10 is shown the principle of a laser triangulation device. Now the geometry of the measurement is chosen so that the output beam of the laser diode is at oblique incidence. However, a small spot of the probe beam is obtained only if the angle of the incidence is near the normal light incidence, which is preferred in this imaging device for the better accuracy in the detection of the position of the light spot in the imaging plane.

The detector is either a position sensitive detector, which is fast, or slower CCD camera. The light spot at the image plane is moving according to the height variation of the surface profile. Once the geometry of the system is known the surface profile is obtained by application of trigonometry. Also here it is crucial to detect high light intensity. Thus the spectral properties of the surface are certainly important. Highly reflecting surfaces are preferred. Remote sensing is possible by furnishing the devise by appropriate optics. For on-line operation, e.g., at cold-rolling line continuous history of the product surface roughness can be documented and, e.g., the change of the roller can be estimated using the triangulation device. The easiest implementation is to fix the position of the measurement head over the metal strip, thus surface profile information along a line, or if there are two or more sensors, along two lines is obtained while the strip is running along the machine direction. A better picture of the surface roughness can be obtained using a scanner unit, which is scanning the strip in the cross machine direction. Totally real time operation of the device is not currently possible, and one cannot get overall surface roughness of the belt with the above described devices. So there is still an order for more efficient surface roughness measurement devices. Nevertheless, quality inspectors in industry are usually happy even with a discrete picture of the surface roughness obtained by an automatic commercial gauge that is monitoring the product condition.

4.5 Low Coherent Proximity Sensor for Surface Roughness Monitoring

Surface roughness and especially its change can be estimated using rather simple and cheap proximity sensors. In Fig. 4.11 is shown one example of such an apparatus [55]. The light source in this particular sensor was a LED operating at the wavelength 941 nm with spectral band width ca. 50 nm. The signal of the sensor depends on the distance from the surface and the surface roughness of the object. In Fig. 4.12a are shown curves for different rms surface roughness of cold-rolled hydro aluminum samples as a function of the measurement distance. Obviously the best sensitivity against the surface roughness can be obtained in the vicinity of the maximum value of the signal. In Fig. 4.12b is presented the signal for a fixed distance of the sensor head from the surface, but for different surface roughness of the aluminum samples this figure suggests a linear response at a relatively narrow surface roughness range.

In the case of a wider surface roughness range the signal tends to be a nonlinear function of the average surface roughness. This is demonstrated in Fig. 4.12c for flat-lapped electroformed nickel surface roughness standards. In the Fig. 4.12d we show data from on-line measurement. The surface roughness signal at the right-hand side was recorded after 100 tons of cold-rolling of the steel was done. The speed of the production line was hundreds of meters per minute. It is obvious that the surface roughness of the strip has become smaller at later instance which can observed if we compare the signals at the left- and right-hand sides in Fig. 4.12d. The reason is that the surface of the roller itself has become smoother. In the insert of Fig. 4.12d is shown how

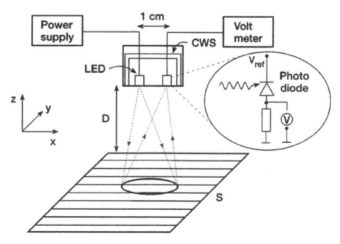

Fig. 4.11. Proximity sensor based on LED. CWS, water cooling system; and S, surface

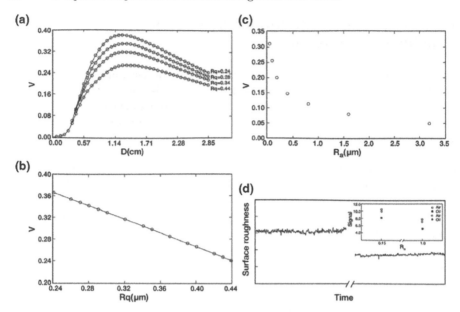

Fig. 4.12. Signals (**a**) and (**b**) from rough aluminum surfaces obtained by proximity sensor, (**c**) signal from nickel surface roughness standards, and (**d**) on-line signal from steel cold-rolling machine (the lower level signal was detected after cold-rolling of 100 tons of steel). In the insert of (**d**) the role of rolling oil on the magnitude of the signal is demonstrated

rolling oil film on the metal has an effect on the signal in the case of two steel samples with different average surface roughness. In cold-rolling process oil is exploited at the roller to reduce friction between the roller and the metal strip. The oil is a disturbance that one has to take care in the case of the optical sensor. Usually one has to pay attention in installing the gauge in a position where external disturbances such as oil, dust and also stray light at the factory are minimized. Vibration of the steel strip is always present but it can be cancelled out by mechanical design of the gauge and signal analysis.

4.6 Low Coherence Interferometer as a Surface Profilometer of Porous Media

Optical coherence tomography (OCT) provides a good means for the measurement of the surface profile from porous samples such as paper, tablets, etc. A practical device is based on the low coherence interferometer which is shown in Fig. 4.13. The light source is a superluminicent diode that produces IR- radiation and has the emission peak power tens of mWs. The diode illuminates

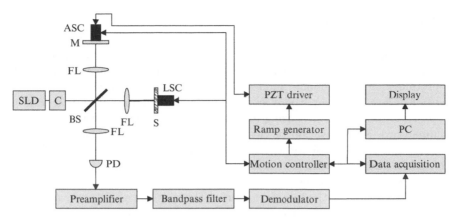

Fig. 4.13. Schematic diagram of OCT devise for measurement of surface roughness of paper

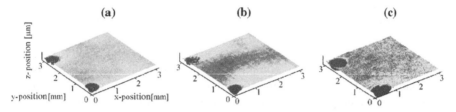

Fig. 4.14. Topography maps from Xerox-, SC- and fine papers recorded by OCT. For positioning of the samples two needle holes were made, which can be distinguished from the pictures. The average surface roughness was 1.78 (fine), 5.40 (SC) and 7.38 μm (Xerox)

a Michelson-type interferometer with a free space configuration. A diverging beam from the superluminiscent diode is collimated and split equally by a cube beam splitter into the two arms of the interferometer. The reference arm contains an axial scanner to produce interference modulation. Scanning can be carried out using a servo motor or a piezoelectric scanner. The object is in the measurement arm, and the object can be scanned, with resolution typically 10–20 μm, for the measurement of the 3D- topography of the surface. When the OCT is used in the low coherence interferometry (LCI) mode one can use the first reflection from the porous sample vs. the depth. In Fig. 4.14 we show topography maps obtained from fine-, super-calendered- and Xerox papers, respectively. Average surface roughness of these three paper grades was obtained, by scanning $3 \times 3\,\mathrm{mm}^2$ area, and making use of (4.1)

Obviously OCT provides a nice device for getting topographic data from a porous media such as paper, pharmaceutical tablets, etc. This device was found useful in the inspection of surface roughness of paper and prints [56,57]

4.7 Specular Gloss

Gloss is an important property in cases where aesthetic impression is required from the product. Such a property is appreciated in connection with multitude of products such as, mobile phones, tableware, draw off pipes, cars, just to mention a few. However, there are cases where negligible gloss is a wanted property. Apparently the sensing of gloss depends heavily on the response of a human eye. The sensation of gloss depends also on the surface roughness, texture and spectroscopic properties of the object. For nonporous media the strength of the light reflection from the surface of an object is crucial for it to be glossy or not. In the case of porous media bulk scattering has to be taken into account in the interpretation of the gloss measurement. Although a human eye is an excellent device for gloss it is not sufficient in sectors of industry, where automation of the production processes is highly appreciated. Hence off- and on-line glossmeters are frequently utilized for routine testing of the product surface. We wish to remark that although gloss has strong emphasis on visual sensation, the definition of gloss is valid throughout the whole electromagnetic spectrum. In a sense gloss can be measured, e.g., in the reflection of terahertz radiation from an object.

Theory and measurement principles on gloss assessment have been described in [58, 59]. An inter comparison between different glossmeters has been also carried out [60]. There are different types of gloss [58, 59], such as specular gloss, contrast gloss, sheen, absence-of-bloom gloss, distinctness-of-reflected-image gloss, and absence-of-surface-texture-gloss, but here we concentrate only on the specular gloss, i.e., the ratio of luminous flux of light reflected from an object to the specular direction to the luminous flux reflected from a gloss standard. From now on we denote the specular gloss simply by the "gloss". The measurement of the gloss has been standardized, e.g., by ASTM and ISO. The measurement principle is simple and it can be extended to off- and on-line measurement of the gloss. The matte or shiny surface is illuminated using a white light source by collimating the beam with a lens as shown in Fig. 4.15. Specularly reflected light is collected by another lens to the photo detector. According to the standard the object has to be planar. A standard reference surface is highly polished black colored glass that has refractive index 1,567. The standard should give 100 specular gloss units (GU) at all angles of incidence, but in principle there is no upper limit of gloss in such a definition. The glossmeters typically make use of the 20, 46, 60, 75 or 85° angle of light incidence. Low angle of incidence is used for shiny and high for dull surfaces, respectively. For example, in the laboratories of paper mills high angle of incidence is utilized in the measurement of the gloss of coated and uncoated paper. Usually a different angle of incidence has to be chosen for any dull or shiny surface even though they are made from the same material.

It is obvious from the previous sections that surface roughness, depending on the topography [61], and finishing marks tend to scatter light incident on an object. Therefore gloss depends on these factors, and in addition to the

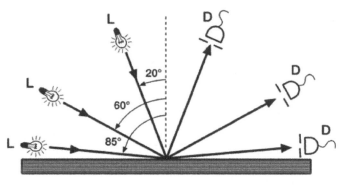

Fig. 4.15. Conventional glossmeter. Typical measurement geometries are illustrated. L, lamp; and D, Detector

spectral properties of the surface, more precisely, the complex refractive index and its spatial variation. Due to the oblique angle of incidence, which is used in gloss measurement, the polarization of the light is also an issue. In the reflection of light the p- and s-polarized electric fields behave in a different manner according to the Fresnel's equations. Actually for rough surfaces the utilization of Fresnel's equations are highly questionable but they provide a qualitative picture to realize that the intensity of the reflected light depends on the polarization of light, angle of incidence and the complex refractive index of the medium. Obviously diffused light from a rough surface is incident also to the specular direction and hence presents a noise, i.e., an error in the measurement of gloss. It is possible that two different surfaces have the same specular reflection but they have different diffuse reflection. Such surfaces can be separated from each other by detection of light scattering either at one or more angles different from the angle of specular reflection. In such cases we speak about contrast gloss.

The requirement of the gloss standards that the object has to be planar is rather restrictive. In commercial glossmeters the probing area of the incident light is usually relatively large. Therefore, inspection of tiny and curved surfaces is usually a problem. Most of the glossmeters require a contact (in paper mills the gloss of supercalendered paper may be detected using on-line noncontact scanning system) with the object thus fragile objects are beyond the scope of inspection with these glossmeters. When it concerns on-line inspection at a process industry where a web is moving, one usually faces the problem of the vertical movement of the web. In Fig. 4.16 we illustrate problems of gloss measurement related to the vertical movement of the web, and also of a curved object.

In Fig. 4.17 we demonstrated problems related to the local gloss variation. Local gloss has importance, e.g., in better understanding of production parameters and in R & D of the quality of the product. For instance, if we consider the print quality on a paper and that the density and gloss mottling have

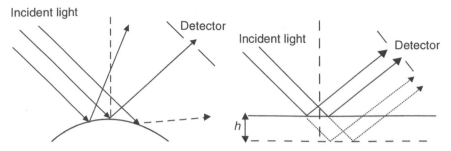

Fig. 4.16. Problematic objects for conventional gloss detection. Curved object (*left*) and vertical movement of a web (*right*)

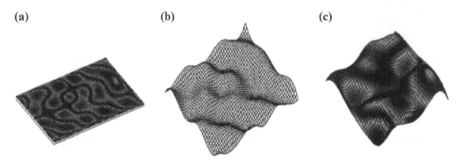

Fig. 4.17. Spatial variation of (**a**) refractive index, (**b**) surface roughness and (**c**) both surface roughness and refractive index make the interpretation of local gloss problematic

importance for the goal of printing better images on books and magazines at the printing houses then readers interested in print gloss and role of the optical properties and surface roughness on the print gloss may consult [62, 63]. The problems of conventional glossmeters can at least be partly overcome using a diffractive optical element based glossmeter, which we describe in Sect. 4.8.

4.8 Diffractive Optical Element Based Glossmeter

Considering that we mostly inspect things so that the line of sight is along the normal of a surface the measurement of gloss at very large angle of light incidence is to some extent artificial. The reason for high angle of gloss measurement is due to the fact that the signal of specular reflection is rather low, especially for matte, rough and porous surfaces. It is also a drawback of gloss measurement at an oblique incidence because of different surface finishes, although dealing with the same material, require different angles of incidence. Hence, it may be difficult to compare the gloss from the same material but having different surface roughness. One solution for this problem is that specular reflection is measured at the normal light incidence. Such a measurement

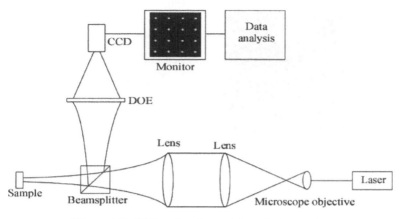

Fig. 4.18. Diffractive element based glossmeter

geometry has also other advantages such as, vertical movement of the product is allowed to some extent and the polarization of the incident light has a negligible role.

A promising measurement device is the one that is based on the diffractive optical element based glossmeter (DOG) [64]. Schematic diagram of the measurement system is shown in Fig. 4.18. This apparatus belongs to the class of machine vision. The light source is a HeNe or a semiconductor laser. By focusing the laser beam and guiding this beam it is possible to measure the local gloss and its variation from tiny and curved objects. In Sect. 3.2 the DOE was used as a projection unit. Here its function is different, namely it is used as an analyzer of scattered light. The incident wave front is distorted due to the roughness of the surface. However, the DOE is sensitive on the amplitude and the phase of the scattered field which is reconstructing the computer-generated hologram. We have compared the gloss obtained by DOG with a conventional commercial glossmeter for flat surfaces. Almost a perfect correlation between the two glossmeter has been found for various materials. This is somehow surprising because the conventional glossmeter uses a white light source, whereas the DOG is based on using monochromatic coherent radiation of a laser. It is interesting to note that the specular reflection is weakest at the normal incidence, nevertheless DOG is rather sensitive at this measurement geometry and it can be utilized practically speaking, to all kind of surfaces, including very dull surfaces, such as, nanocarbon surfaces.

In the case of the DOG the gloss is defined with the aid of image data shown in Fig. 4.19. The total intensity of image pattern is calculated inside the area, marked by a dashed line in Fig. 4.19, as follows:

$$I = \frac{1}{nm} \sum_{i=1}^{n} \sum_{j=1}^{m} I_{i,j}, \tag{4.16}$$

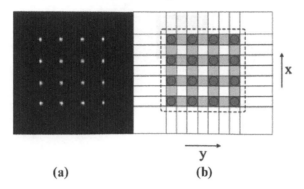

(a)   (b)

Fig. 4.19. Image data of DOG used in gloss analysis

where n and m are the dimensions of the DOE image and $I_{i,j}$ is the image intensity at the (i,j)th element of the CCD camera detector array. The gloss is defined as follows:

$$G = \frac{I_S}{I_R} \times 100, \tag{4.17}$$

where I_S and I_R are measured from the sample and the reference, which is a high quality mirror or a commercial gloss standard. A practical measure of the surface anisotropy, such as orientation of the finishing marks, is the visibility of the 16 light spots matrix given by

$$V = \frac{I_{\mathrm{Max}} - I_{\mathrm{Min}}}{I_{\mathrm{Max}} + I_{\mathrm{Min}}}, \tag{4.18}$$

where I_{Max} is the mean of the maxima of the 16 peaks and I_{Min} is the mean of the minima between the peaks. Visibility is obtained both at the x- and y-directions.

In Fig. 4.20 (right) is presented an injection mold black colour plastic product, made from acrylonitrile-butadiene-styrene (ABS) in polycarbonate blend (PC), and three locations a, diffuse; b, glossy; and c, semi-glossy, where the gloss was measured. The change of the local surface roughness is due to the local wear of the tool, which is increasing as it is a function of the duty cycle of the tool. The waist of the laser beam was 30 μm and a translation stage was used in scanning the positions a, b and c. According to the data, which is shown in Fig. 4.20 (left), a linear relationship between the decrease of surface roughness (wear of the tool) and gloss could be recorded for the plastic products, that were taken at various stages of the production process.

DOG has been applied also in inspection of glazed ceramics, such as interior of coffee cups, and the wear of the tableware due to machine washing. Since DOG provides information on the local gloss one gets, by scanning the laser beam or the sample, as also two-dimensional gloss map of the surface one example of such a map is presented in Fig. 4.21 for a paper sample.

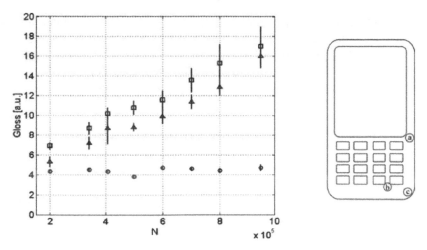

Fig. 4.20. Gloss evolvement of a plastic product as a function of duty cycle of the molding tool. Location a (*circle*), location b (*triangle*) and location c (*square*)

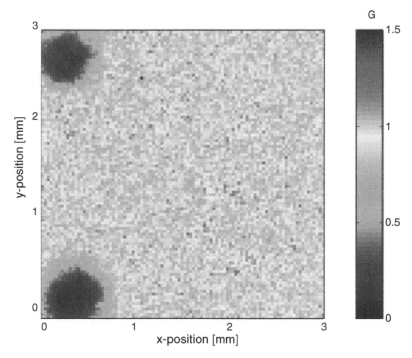

Fig. 4.21. Color-coded gloss map of a paper sample. The circular patterns at the left corners are needle holes, which can be used in positioning of the sample. Local variation of the gloss in this figure is due to the local surface roughness and refractive index of the paper

By DOG inspection of metals, which were produced from different steel alloys and by different finishing processes and which are used in plastics injection molding, these were examined with the aim to get better surface quality than that of the highest standard surface. This was successful, and information about the anisotropy of surface marks could be also detected [65].

4.9 Statistical Parameters for Gloss Assessment

In Sect. 4.8 we introduced a point wise scan of the specular gloss using the DOG. Next we present how the data of DOG can be analyzed using statistical models that are analogous to those we already have used for the description of the surface roughness. We follow the definitions presented in [66]. First we define average and rms gloss variation as follows:

$$G_{\mathrm{a}} = \frac{1}{A} \iint_A |G(x,y) - \langle G(x,y) \rangle| \mathrm{d}x \, \mathrm{d}y, \qquad (4.19)$$

and

$$G_{\mathrm{q}} = \left\{ \frac{1}{A} \iint_A [G(x,y) - \langle G(x,y) \rangle]^2 \mathrm{d}x \, \mathrm{d}y \right\}^{1/2}, \qquad (4.20)$$

where $G = G(x,y)$ is the gloss as a function of location and $\langle G(x,y) \rangle$ is a mean gloss in a manner that $G(x,y)$ has a minimum variance. As a special case, namely gloss parameters along a straight line is a simplification of (4.19) and (4.20) into one dimension.

Next we can define slope parameters of gloss using the concept of partial derivatives as follows:

$$G_{\mathrm{a},x} = \frac{1}{A} \iint_A \left| \frac{\partial G(x,y)}{\partial x} - \frac{\partial \langle G(x,y) \rangle}{\partial x} \right| \mathrm{d}x \, \mathrm{d}y$$

$$G_{\mathrm{a},y} = \frac{1}{A} \iint_A \left| \frac{\partial G(x,y)}{\partial y} - \frac{\partial \langle G(x,y) \rangle}{\partial y} \right| \mathrm{d}x \, \mathrm{d}y \qquad (4.21)$$

and

$$G_{\mathrm{q},x} = \left\{ \frac{1}{A} \iint_A \left[\frac{\partial G(x,y)}{\partial x} - \frac{\partial \langle G(x,y) \rangle}{\partial x} \right]^2 \mathrm{d}x \, \mathrm{d}y \right\}^{1/2}$$

$$G_{\mathrm{q},y} = \left\{ \frac{1}{A} \iint_A \left[\frac{\partial G(x,y)}{\partial y} - \frac{\partial \langle G(x,y) \rangle}{\partial y} \right]^2 \mathrm{d}x \, \mathrm{d}y \right\}^{1/2}. \qquad (4.22)$$

In general case the slope has to be defined with the aid of using directional derivatives using the concept of a gradient.

Other useful parameters are the autocorrelation (AC) and power spectral density function (PDF) of gloss. The autocorrelation function quantifies the similarities of the gloss profile in lateral direction, whereas PDF is obtained by squared modulus of Fourier transformation, describes gloss fluctuations, especially their periodicity in spatial frequency plane (f_1, f_2). These functions are defined as follows:

$$AC(\tau_1, \tau_2) = \frac{1}{A} \iint\limits_A G(x,y)z(x+\tau_1, y+\tau_2)\mathrm{d}x\,\mathrm{d}y, \qquad (4.23)$$

$$PDF(f_1, f_2) = \frac{1}{A} \left| \iint\limits_A G(x,y)\mathrm{e}^{2\pi i(xf_1+yf_2)}\mathrm{d}x\,\mathrm{d}y \right|^2. \qquad (4.24)$$

The most convenient way usually is it to calculate AC and PDF for the one-dimensional case in surface roughness studies. Thus in the analogous way the autocorrelation length of gloss, $l = AC(0)/10$, is defined as a measure on how much the one-dimensional gloss profile resembles itself. The gloss profile is obtained by sampling discrete but equispaced data points. Thus in data analysis one has to utilize approximations of the above mentioned parameters by replacing the integrals involved with corresponding sums.

Despite the resemblance between the definitions of the surface roughness and gloss parameters, there are crucial differences. The gloss depends not only on the surface roughness and texture of finishing marks, which are due to surface processing, but also on the complex refractive index of the medium. In Fig. 4.17 we have illustrated different kind of surfaces where the complexity of factors for gloss is demonstrated. Nevertheless, the statistical

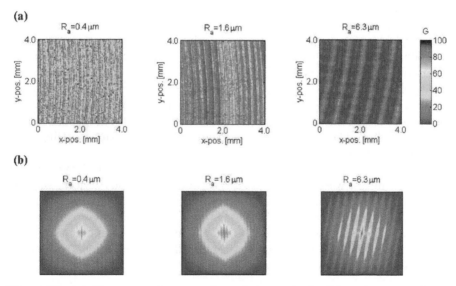

Fig. 4.22. (a) Gloss maps from metals surface standards, (b) and autocorrelation

Table 4.1. Statistical gloss parameters for some metal surface roughness standards

	$R_a = 0.4\,\mu m$	$R_a = 1.6\,\mu m$	$R_a = 6.3\,\mu m$
G_{mean}	34.85	17.89	5.60
G_a	16.15	8.00	2.61
G_q	20.13	9.66	3.58
$G_{a,x}$	7.40	3.32	1.08
$G_{a,y}$	3.97	1.53	0.97
$G_{q,x}$	9.76	4.40	1.56
$G_{q,y}$	5.38	2.11	1.38

parameters for the gloss, given above, are general and cover also complex surface structures. In Fig. 4.22a) we show gloss maps for metals surface roughness standards (turned surfaces) and in Fig. 4.22b the corresponding AC patterns. In Table 4.1 are calculated statistical gloss parameters for these samples. The statistical gloss parameters help to estimate the success of the finishing process of the surface, and also its quality. Obviously, statistical gloss analysis has applications in various sectors of industry. The data of Fig. 4.22 can be used in estimation of the spatial features of gloss and mottling of the products. The periodicity of the finishing marks is obvious from the data of Fig. 4.22.

Oksman et al. have provided a sensor [67] and statistical parameters [68] for the evaluation of contrast gloss, which together with specular gloss yield a wider picture about the phenomenon that is understood as a "gloss".

4.10 Conclusions

Inspection of surface quality has been and will be an issue in various industrial sectors such as metal, paper, plastic, etc. to mention a few examples. Nowadays a typical surface inspection involves a point or line sensing of the product. Unfortunately, such a measurement technique will not provide comprehensive information of the whole product. Therefore, systems that can measure a large area quickly are favored in industrial environments. There are measurements that are carried out even now in laboratory, such as gloss of a print. The trend will be, for example in the case of printing houses, that print quality can be inspected on-line. There the issues are dynamic print and print mottling with sufficient statistical analysis of the measured data.

It is much more difficult task to carry out an automatic inspection of surface roughness, waviness and gloss of a complex shaped object, such as the interior of a coffee cup, or the tool needed in injection molding of plastic covers of mobile telephone and car accessories. These challenges can be met by optical noncontact measurement techniques but research and development is surely needed before appropriate, preferentially multimeasurement gauges, are brought into the markets.

Measurement of Positions, Distances, and Displacement

Sensing of light beam length (L, distance) and position (x,y) has a lot of applications. Derivation of the continuous distance value gives the speed of an object. Continuous distance measurement determines also the displacement towards the observer (ΔL). Beam position measurement can be used to measure vertical (Δy) or horizontal (Δx) displacement and angle of the beam (α) or angular displacement ($\Delta \alpha$) as given in Fig. 5.1.

Lasers and light emitting diodes (LED) as a light source can effectively be used to measure distance, displacement, or position of an object. In particular, noncontacting distance measurements to a noncooperative target are of great interests for many industrial, sport, and traffic control applications.

Position sensitive detector (PSD) is an effective tool to measure the position of a light beam (Fig. 5.1). With appropriative optical construction, it can also be used to determine the direction to light source or to an illuminated light spot on diffusively reflecting surface, enabling, e.g., distance measurement or tracking of a moving object. Another variation of the measurement is to use a reflector on the object surface and wide beam illumination, Fig. 5.2.

As depicted in Fig. 5.2 the cooperative target point is illuminated with an infrared light cone, whether a laser diode (LD) or LED. A part of the light reflected from the reflector is collected and focused by the receiver lens on the PSD. Its output is directly proportional to the displacement of the light spot from its centre, which is in turn directly proportional to the angular displacement of the target reflector from the optical axis of the receiver. The angular displacement for small angles is

$$\alpha \approx \Delta y / f_{\mathrm{e}}, \tag{5.1}$$

where Δy is the displacement of the image of the reflector or light spot from the centre of the PSD and f_{e} is the effective focal length of the receiver lens [69].

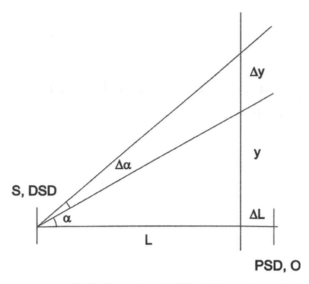

Fig. 5.1. Measurement of light beam length (L, ΔL), position (y, Δy), and angle (α, $\Delta \alpha$). S, DSD is light source or distance sensing device and PSD, O is position sensitive detector or object

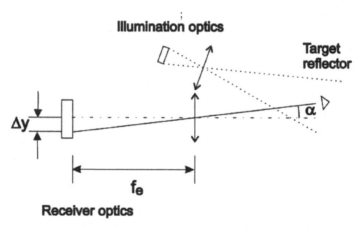

Fig. 5.2. Position sensing of a reflector using wide beam illumination and PSD

5.1 Distance Measurement

Good reviews of absolute distance measurement can be found, e.g., in [70,71]. Latest achievements are provided in Proceedings of ODIMAP (Optoelectronic distance/displacement measurements and applications) Series of Topical Meetings, held in Nantes, France (1997), in Pavia, Italy (1999 and 2001), in Oulu, Finland (2004), and in Madrid, Spain (2006).

Widespread main techniques used by laser distance meters are triangulation, interferometers, and time-of-flight described in the following:

- *Triangulation* is basically a geometric method, useful for distances in the range from 1 mm to many kilometres. It measures the angle (α) from two points to target. Distance between the points is called base (D). Distance to target (L) is found as follows: [72]

$$L = D/\tan \alpha \approx D/\alpha. \tag{5.2}$$

Figure 5.3 shows schematics of a laser triangulation probe for displacement (ΔL) measurement constructed using PSD as detector for $\Delta\alpha$.

A laser projects a spot of light on a diffuse surface of an object, and a lens collects a part of the light scattered from this surface to image the spot on a position sensor. If the object is displaced from its original position by a small amount (ΔL), the center of the image spot will also be displaced by an amount Δy from its original position. Therefore, the displacement of this object can be determined by measuring the displacement of the image spot centre on the position sensor. A laser triangulation usually measures the displacement of an object in longitudinal direction (e.g., along laser beam). Based on (5.2) the displacement can be calculated as given in the following formula:

$$\Delta L = -(D/\alpha^2)\Delta\alpha = -(L^2/D)\Delta\alpha. \tag{5.3}$$

Equation (5.3) can also be used to evaluate the relative error of distance measurement if $\Delta\alpha$ is an error of the angle measurement, as follows:

$$\Delta L/L = -(L/D)\Delta\alpha. \tag{5.4}$$

The accuracy is typically $L/1\ 000 \dots 10\ 000$, if D and L are of the same order. The accuracy is mainly determined by the PSD resolution and linearity.

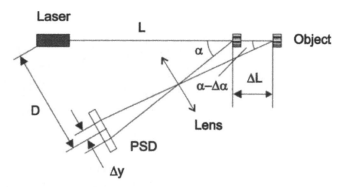

Fig. 5.3. Schematics of the conventional laser triangulation method for distance or displacement measurement

– *Interferometers* allow distance measurements with accuracy related to the wavelength of the light source. Usually a coherent beam is needed and the object should be mirror-like. This method can cause so-called unambiguous problem to be solved. In some cases white light as a light source can be used. A great advantage of white-light interferometer is that it can be used to measure objects with rough surface and because of its theoretically unlimited unambiguous range [73]. Figure 5.4a shows a scheme of Michelson interferometer which can be used with coherent and wide band light source.

Distance resolution (ΔL) of the white light interferometer is determined by the coherence length of the light source. It can be calculated from the bandwidth ($\delta\lambda$) of the light source as follows [74]:

$$L_c = \Delta L = \ln(2)(2/\pi)(\lambda_0^2/\Delta\lambda), \qquad (5.5)$$

where λ_0 is mean wavelength of the light source. The coherence length in (5.5) is the round-trip coherence length. The equation is valid in vacuum. The distances measured by white light interferometer are optical distances. This means that the geometric distance is derived by dividing the optical distance by the refractive index (n) of the media. The coherence lengths of the light sources in question are typically 1–10 µm, which limits the minimum thickness of the measured object or separation between reflecting details. The method is suitable to measure surface profile, thickness of a foil or a coating, for example. The method can be used also for scattering material and tomography. By adding the scanning, 3D measurements can be realized. It is called as optical coherence tomography (OCT) and be used, e.g., to analyze human tissue [75] or paper structure [76].

Figure 5.4 shows also the output signals as a function of time if the reference mirror is moving with constant speed. The fringe pattern frequency in time domain is the Doppler shift of the light reflected from the moving mirror.

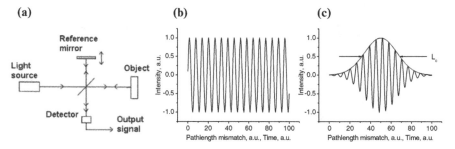

(a)   **(b)**   **(c)**

Fig. 5.4. (a) Michelson interferometer, (b) interferogram for coherent light and (c) for white light source as a function of reference mirror movement (pathlength mismatch). L_c is coherence length of the light source and determines the FWHM (full width at half maximum) of the envelope of the interferogram

It can be useful for electronics design because narrow band detection can be used to maximize the *S/N* (signal-to-noise ratio).

For conventional Michelson interferometer, which use coherent light source, the resolution is related to phase measurement accuracy inside a cycle being even at subnanometer level, but the technique can be used only for mirror-like surfaces and for limited distance range and to measure displacement, not absolute distance.

- *Time-of-flight* (TOF) method refers to the time it takes for the emitted light to travel from its transmitter to an observed object and then back to the receiver (t_d). Distance to the target is found as follows:

$$L = ct_d/2. \qquad (5.6)$$

Speed of the light (c) is roughly 30 cm ns^{-1}. The method may usually be used for distances more than 1 m and accuracy is principally not dependent on distance. Typical accuracies of simple devices for short distances are a few millimeters or centimeters. Time-of-flight method can be realized by many different methods. Mostly used principles are: phase shift method [77], frequency modulated continuous wave (FMCW) method [78] and pulsed TOF method [79].

- *Phase shift or phase difference method* is a common technology for surveying. It allows to measure distances as depicted in Fig. 5.5.

A sinusoidally modulated laser beam is sent to a target, usually a reflector, e.g., a corner cube. Reflected light (from diffuse or specular reflections) is detected, and the phase of the power modulation is compared with that of

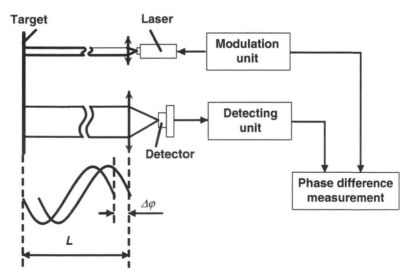

Fig. 5.5. Phase-shift method for distance measurement

the sent light. The obtained phase shift is 2π times the time-of-flight ($2L/c$) times the modulation frequency (f), but the number of full cycles elapsed during the flight of a specific signal has to be taken into account to solve the ambiguity problem. If the total phase shift ($\Delta\varphi$) is inside one cycle, it leads to a distance as follows:

$$L = \Delta\varphi c/4\pi f. \tag{5.7}$$

By choosing an appropriate modulation frequency, one can adjust the distance sensitivity to the required level. Considering as a reasonable hypothesis a phase measurement resolution of $0.1°$, it is possible to measure displacement $\Delta L = 2\,\text{mm}$ with $f = 20\,\text{MHz}$ and maximum unambiguous measurement range of $7.5\,\text{m}$ ($c/2f$) [80].

- *FMCW-technique* is called also as pulse compression, chirped frequency or ramped frequency modulation. This method is a subcategory of the pulsed time-of-flight method which combines the high energy of long pulse with the high resolution of a short pulse. An example of the pulse structure is shown in Fig. 5.6.

Compared to continuous wave phase shift method in this case the frequency of the sinusoidal modulated pulsed beam will be modulated by some function. This function can be triangular, saw tooth (ramp) or sinusoidal in shape. The distance is a function of the frequency shift (Δf) between the transmitted and received signal as follows (Fig. 5.6):

$$L = \Delta f c T/2 f_a. \tag{5.8}$$

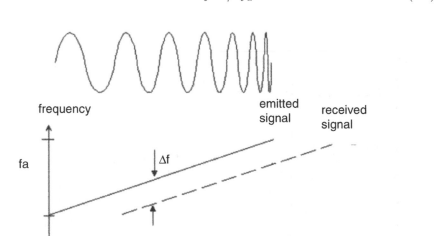

Fig. 5.6. Transmitted signal of FMCW distance measurement, f_a is the frequency shift during the time of one sweep (T), and Δf is the frequency difference between the transmitted and received signal and t_d is the pulse transit time

Usually matched filter detection is used. The filter can be realized by using swept frequency filter or delay lines. The measurement resolution is inversely proportional to the sweeping range of the modulation frequency. The ability of the receiver to improve the range resolution over that of the conventional system (pulse length) is called the pulse compression ratio $G = f_a T$. It can be used to calculate the distance resolution as follows [81]:

$$\Delta L = cT/2G = c/2f_a. \tag{5.9}$$

Alternatively, G gives also the reduction ratio of the peak power needed to get the resolution without compression. The minimum range is not improved by the process. As in the case of microwave radars also the optical frequency (wavelength) of the laser can be modulated. For a tuning range of $\sim 250\,\text{GHz}$ a measurement accuracy of about $0.2\,\text{mm}$ can be achieved. Dynamical effects such as movement of the object of interest or scanning of the laser beam can cause large systematic measurement errors due to the Doppler shift or time-dependent phase variations.

Pulse compression can also be realized using envelope modulation of the carrier signal by complementary binary codes like Barker- or Frank-codes [82]. The receiver is of correlation type calculating continuously the cross correlation between the transmitted and received pulses. The delay at correlation maximum gives the time-of-flight. The compression ratio is now the length of the code used.

Optical distance measurement is very broad and fast growing discipline. This chapter concentrates on nonmodulated, pulsed time-of-flight distance measurement providing also some promising applications.

5.1.1 Laser Pulse Time-of-Flight Distance Measurement

The pulsed laser distance meters are often called LADARs or LIDARs (LADAR, laser detection and ranging and LIDAR, light detection and ranging) or with the term laser range finder. The term laser radar includes both laser range finders and the devices measuring 3D-distance picture of the object or the absorption and scattering of light from the atmosphere. In this book the term laser radar will be used for 3D distance measurement.

A TOF system measures the round trip time between a light pulse emission and the return of the pulse echo from the object. Using elementary physics, distance is determined by multiplying the velocity of light by the time light takes to travel the distance. In this case the measured time is representative of travelling twice the distance and must, therefore, be reduced by half to give the actual range to the target (5.4). To obtain $1\,\text{mm}$ accuracy, the accuracy of the time interval measurement should be $6.7\,\text{ps}$ [71].

Since a single pulse is adequate for the unequivocal determination of distance with centimeter precision and accuracy depends only weakly on distance, this method is particularly appropriate, for example, in applications involving distances longer than 1 m, in applications where reflectors are not used and in fast measurement applications such as scanning or measuring moving objects. In addition, averaging enables millimeter or even sub millimeter precision to be achieved. An additional advantage of the pulsed TOF system arises from the direct nature of its sensing as both the transmitted and returned signals follow essentially the same direct path to an object and back to the receiver minimizing the shadowing effect in measuring complicated surfaces.

Some new applications such as sensors for traffic control, vehicles with anti-collision alarm and proximity, and sensors used in protecting an area in front of a machine (safety guard) demand limitations on instrument size, mass, and power consumption. To achieve these goals, the basic building blocks of a TOF range finder have to be realized in the form of high-performance integrated circuits.

Constructions

A pulsed TOF distance measuring device (range finder) consists of a laser transmitter emitting pulses with a duration of 1–50 ns, a receiver channel including a PIN or an avalanche photodiode (APD), amplifiers, an AGC (automatic gain control), timing discriminators and time interval measurement block. The emitted light pulse (start pulse) triggers the time interval measurement unit, and the reflected light pulse (stop pulse) stops it. The distance to the target is proportional to the time interval. A block diagram of a laser range finder is shown in Fig. 5.7.

The selection of laser type depends on the intended measurement range and the required speed. For long distances (up to several kilometres), a

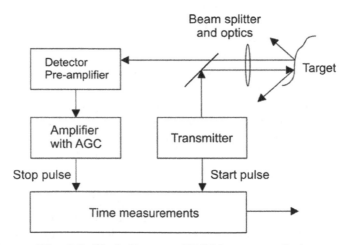

Fig. 5.7. Block diagram of TOF laser range finder

Q-switched Nd-YAG laser or fiber laser can be used, giving peak powers extending to the megawatt level. Low-priced pulsed laser diodes, capable of producing peak powers of tens of watts, enable measuring distances up to a few hundreds of meters – or even longer using coherent summing or pulse compression. The repetition frequency of YAG lasers is low, whereas laser diodes can be used at rates of tens of kilohertz, the DH-type may even reach the megahertz level. Diode pumped YAG lasers and fiber lasers are developing fast and will find applications for range finding. To get eye-safety system instead of silicon based components (400–1,100 nm wavelength) InGaAS-components offer a solution for that (1,500 nm wavelength operation).

Amplification stages must be highly linear and sensitive (high S/N), and have stabilized delay with sufficient bandwidth to follow input signal. They consist typically of preamplifiers, attenuators, and postamplifiers. The preamplifier converts the photodiode current into voltage and is typically of transimpedance type. The main advantage of a transimpedance amplifier is the low input impedance. The higher the value of resistor is the higher is signal-to-noise ratio but the bandwidth is lower. The postamplifier is typically a voltage amplifier, and may be gain controlled.

The dynamic range of the optical input signal often exceeds the input dynamic range of the timing discriminator (time pick-off circuit) of the time interval measurement block, because of the variation in the distance from the target and its reflectivity. Signal also has to be reduced to a level with minimum walk error of timing discriminator. In order to overcome this issue, several types of attenuators can be used at amplification stage. The attenuators can be realized for instance by current mode gain control cell, by R–2R ladders or by adjusting APD bias voltage [83]. An adjustable optical attenuator at the receiver optics can be used to realize gain control. The advantage of this method over electrical gain control is its delay stability over a wide control range but may be too slow for many applications. Electronic attenuators should be placed after the amplifiers for noise minimization but before for linearity improvement. Attenuation is defined according to signal amplitude which is determined by a peak detector.

The transmitter and time-interval measurement units are also critical for the accuracy of the system. The transmitter should be able to produce a stable laser pulse shape. This may call for the temperature stabilization of the diode. The dynamic phenomena of the laser diodes, relaxation oscillations, should also be considered when designing the pulsing scheme as they may easily lead to significant changes in laser pulse shape.

The time interval between the start and stop pulses is measured with the time-to-digital converter (TDC), which is a fast, accurate, and stable time-interval measuring device that uses, e.g., a digital counting technique together with an analogue or digital interpolation method [84]. The single shot resolution of the TDC is typically better than the noise generated timing jitter.

Timing Jitter and Walk, Nonlinearity, Drift, and Other Error Sources

The main sources of inaccuracy in laser range finders are noise-generated timing jitter, walk, nonlinearity, and drift. Typical noise sources include noise generated by the electronics, shot noise caused by the background radiation-induced current and shot noise created by the noise of the signal current. Jitter in timing determines mainly the precision of the range measurement. The amount of timing jitter (σ_t) is proportional to noise amplitude (σ_u) and inversely proportional to the slope of the timing pulse at the moment of timing (du/dt). It can be approximated with the triangular rule:

$$\sigma_t = (\sigma_u)/(du/dt). \tag{5.10}$$

A single-shot resolution of 1 cm can typically be achieved with a good signal to-noise-ratio ($S/N = 100$) using the 100 MHz bandwidth of the receiver channel. However, precision deteriorates as the distance increases and the pulse amplitude decreases proportional to the square of the distance. Pulse amplitude and shape variations create timing error in the time-pick-off circuit and that error is called walk error. Jitter and walk in leading edge timing are shown in Fig. 5.8 [85].

The time discriminator is a very important part of a precision time measurement system. The task of the discriminator is to observe time information from the electric pulse of the detector preamplifier and to produce a triggering signal at the right instant. The choice of time derivation method depends on the desired time resolution, counting rate and required dynamic range of the pulse. Commonly used principles in discriminator design include leading edge timing (constant amplitude), zero crossing timing (derivation), first moment timing (integration), and constant fraction timing. Constant fraction discrimination (CFD) compensates with idealized pulse shapes for walk

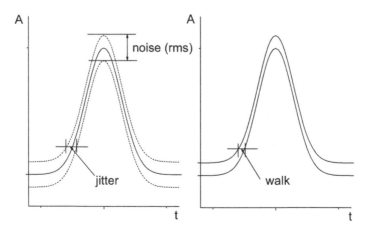

Fig. 5.8. Timing jitter and walk

caused by both amplitude and rise time variations and is commonly used in the time-of-flight measuring units of laser range finders.

The principle behind the operation of CFD is the search for an instant in the pulse when its height bears a constant ratio to pulse amplitude. The occurrence of this point produces a triggering pulse. The constant fraction instant can be realized using a delay cable and comparator so that an attenuated signal coming from the receiver is fed directly to the negative input and the delayed signal to the positive input of an ultrafast emitter coupled logic (ECL) comparator, which amplifies the difference of the attenuated and the delayed pulses [86]. From the output of comparator the signal is fed to the TDC. CFD compensates for walk caused by amplitude and rise-time, but not for walk caused by nonlinear shape variations. Zero crossing and first moment timing compensate for amplitude variations, while leading edge timing fails to compensate for any variation described.

The final precision of the distance measurement can be greatly improved by averaging, with the improvement being proportional to the square root of the number of results averaged. Thus, by averaging 100 successive measurements, the final resolution can be improved to the millimeter level, the corresponding measurement time being 1 ms with a pulsing rate of 100 kHz, for instance.

If the statistical error is averaged to a negligible level, the accuracy of the system is defined by its systematic errors such as nonlinearity in the time interval measurement scale and drift. A careful design of the system can reduce these errors to the millimeter level.

The electronics performance must be based on the specified technical requirements, e.g., range accuracy and power consumption. The main factors limiting the measurement accuracy of TOF laser range finder can be concluded as follows:

(1) Noise, which is the main random error and defines receiver precision. It consists mainly of dark current and multiplication noise from the APD, noise from amplifiers, attenuators and background light.
(2) Walk error, based on pulse amplitude and shape variation and thus on changes their timing point, on nonlinearity and on nonideal gain control changing propagation time of the pulses, causing systematic errors.
(3) Linearity of almost all components: APD, pre/postamplifiers, attenuators and time interval measurement block.
(4) Stability, mainly influenced by device temperature variations.
(5) Electronics bandwidth which has to be high enough for the receiver optical signal, i.e., electronics bandwidth has to include the input signal spectrum.
(6) Dynamic range with minimum limit defined mainly by electronics noise and upper limit by APD maximum photocurrent, and by amplifiers operating voltage range.
(7) Based on finite beam diameter many error sources appear. Inside illuminated area reflectivity variation, object shape, multiple objects, etc. may cause erroneous result or open to various interpretations.

Novel Applications and Development Trends

Several new applications for the transit time measurement of laser pulses are currently being developed to supplement traditional ones. A promising method involves using the path length of light pulses in human tissue, pulp and paper, or optical fiber as a sensor principle. In this context, the term photon migration is often used to describe the propagation of light in scattering media like human tissue. In turbid media, photons take a number of different paths thereby broadening short light pulses. The use of time-of-flight techniques for imaging soft tissues (optical tomography) is being actively investigated by many researchers today for applications such as breast cancer diagnosis and imaging the oxygenation state of the brain in new-born infants [87], see Sect. 5.5.2.

From the point of view of the paper industry, one of the most important properties of graphic paper is the interaction of light and paper structure. One method of obtaining a more detailed understanding of the propagation of light in paper and pulp is the high resolution measurements of the delay of the light pulses suffer as they pass through a sheet of paper or a sample of pulp [88], see Sect. 5.5.3.

As for composite materials, optical fibers can be embedded in them during the manufacturing process. Time-of-flight of the light pulse in optical fiber can be measured with a resolution of few picoseconds using similar technology as OTDR (optical time domain reflectometer). Calculations are then based on the fact that the time-of-flight of a light pulse in a fiber is a function of the length and refractive index of the fiber, which are affected by stress, temperature, and pressure [89], see Sect. 5.5.4.

Commercial applications in the civilian sector place several demands on laser-based devices. Firstly, the so-called Class 1 laser condition (eye-safety) that limits the peak power of the pulsed laser to a few watts should be fulfilled. However, the reliable detection of low-reflectance or high-temperature targets requires a sufficiently high optical peak power. The use of picosecond pulses helps to overcome the eye-safety problem, but at the cost of a receiver with high bandwidth. Shorter pulses also give better precision [90]. Laser light in the far infrared (over 1,400 nm) spectrum is called eye safe if moderate power levels are used and it is a trend to move to that wavelength area for applications to be used everywhere. The extent of ocular damage is determined by the laser irradiance, exposure duration, and beam size.

Maximum permissible exposure (MPE), is the level of laser radiation to which a person may be exposed without hazardous effects or biological changes in the eye. MPE levels are determined as a function of laser wavelength, exposure time, and pulse repetition. The MPE is usually expressed either in terms of radiant exposure in $J \, cm^{-2}$ or as irradiance in $W \, cm^{-2}$ for a given wavelength and exposure duration. These limits are determined by international laser standards.

(a)   **(b)**   **(c)**

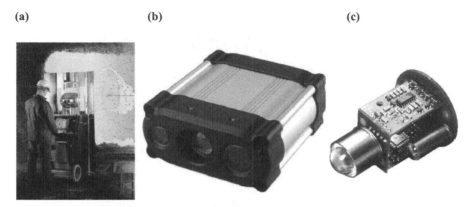

Fig. 5.9. Three development phases of pulsed TOF rangefinder research at the University of Oulu, Finland and Noptel Oy (Courtesy of Noptel)

Other highly desirable qualities of the range finder include cost-effectiveness. The size, weight, and power consumption of the device should be reduced to increase its potential application range. These goals can be achieved using full custom application-specific integrated circuits (ASIC). A long-term vision could be to realize the TOF range finder as a component-like microsystem where all the basic elements (laser diode, photodetector, receiver channel and time interval measurement electronics) are located on a single encapsulated hybrid circuit, Fig. 5.9.

The laser radar in Fig. 5.9a is a commercial product for the measurement of the hot wear thickness of converter in steel factories. It has been constructed using discrete components and achieves cm-level measurement accuracy with a power consumption of about 20 W. The x, y scanning is provided manually. A small range finder for traffic control applications, shown in Fig. 5.9b, includes range finder with speed measurement capability, utilizes ASICs in the receiver channel and in time interval measurement functions. Figure 5.9c show an example of a compact range finder for harsh environment.

5.2 Laser Radar

In 3D measurements, the laser range finder is equipped with angle encoders to enable the definition of the co-ordinates of the measurement point. Scanning is mechanical and is carried out manually or automatically. In some applications, manual scanning is adequate (Fig. 5.9a) but for time-critical purposes a servo system is needed to increase the measurement rate. Basic techniques include scanning either the measuring head or only the measuring beam by means of galvanometer-driven mirrors. Laser radar is a device which uses one of the distance measurement techniques as described before, and scans the direction

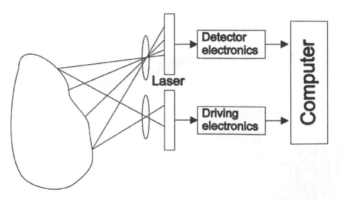

Fig. 5.10. Block diagram of the focal plane scanning

of the distance measurement in two dimensions. This allows to generate a distance image, or more precisely a depth profile of some object, as required, e.g., in robotics.

Focal plane scanning can be used instead of a narrow laser beam which scans a surface mechanically, point by point. This allows a range map to be obtained without mechanical beam scanning. The result is highly improved 3D mapping performance, particularly with respect to measuring time, at considerably reduced mechanical complexity and thus reduced size and power requirements. The principle of focal plane scanning is presented in Fig. 5.10.

One laser beam illuminates its field of view on the surface. The illuminated part of surface is viewed using matrices of separate detectors, usually an APD array. Each detector covers its own fraction of the field of view illuminated by the laser. Detector signals are analyzed in the time domain, and distances to particular points are calculated on the basis of time interval measurements. The system can simultaneously measure distances to several directions without any moving parts [91].

The APD arrays are developing fast also based on the research for PET (positron emission tomography) scintillation detectors to replace photomultiplier tubes [92]. Receiver APD arrays have typically from 2×2 up to 126×126 (still in development) pixels. Most of them do not include integrated circuitry for time interval measurement. One exception is the 32×32 APD array from MIT Lincoln Laboratory with $0.35\,\mu m$ CMOS digital timing circuits at each pixel. Each hybrid pixel then contains a Geiger-mode APD and a timing circuit [93, 94].

These 32×32 APD arrays have been fabricated with single diode active area of $30\,\mu m$ diameter. The pitch on the APD arrays is $100\,\mu m$, yielding 7% fill factor, which is very low compared with the $\sim 90\%$ pretended to the three potential applications (rendezvous and docking, planetary landing and rover navigation). While it is possible to increase the fill factor by bonding a micro lens array to the detector array, this would also increase the background light flux. An alternative is the use of a diffractive optical element to transform the single-mode laser beam into a 32×32 spot array pattern to the target. Optics must be aligned so that each laser spot is matched to an individual pixel. This technique increases the fill factor and also reduces background false detections.

For acquiring such depth profiles at a higher rate, there are sensor chips similar to CCDs (charge coupled devices) with internal electronics to detect phase shifts, so that the distance for each pixel can be measured simultaneously. This allows for rapid three-dimensional imaging with very compact devices [95].

5.3 Gated Imaging

Range gated imaging systems use pulsed laser illumination to form an image of the target. In such systems the exposure time of the camera is synchronized with the arrival time of the transmitted laser pulse so that the imager detects light coming from a predetermined distance only, thus eliminating backscattering and unwanted reflections outside the range of interest. The main advantages of range-gated cameras over passive ones include simultaneous ranging capability and capability to see through smoke, fog and other obscurants such as vegetation. Compared to thermal IR or microwave radar imagers they provide poorer weather penetration but better angular resolution due to short wavelengths used for illumination. Besides in military applications gated imaging systems are used in search and rescue missions, and they have great potential in the field of vehicle enhanced vision. The principle of gated viewing is illustrated in Fig. 5.11.

A short laser pulse is illuminating the scene and a high gain image intensifier tube is gain controlled in time (gated) so that the result is series of images in range slices. The gate delay (τ) is synchronized to the transmitted laser pulse transit time and the gate width (τ_g) to the laser pulse duration. By sliding with the gate one can obtain 3D imagery from a series of 2D images. The tubes which can be sensitive at eye safe wavelengths ($1.5\,\mu$m) have high spatial resolution (30–$60\,$lp mm^{-1}) so that long range target recognition can be made using proper optics. Alternatively, enhanced CCD cameras can capture the pictures as a function of laser pulse delay [96]. If sufficient laser pulse energy is used one can see through fog and other shields [97].

Typically very short ($<1\,$ns) and powerful (up to $1\,$MW) laser pulses are needed in range gated systems. The major issues of illumination pulses are total power, pulse width, wavelength, and jitter. The total power limits the operating range and the scattering density for which the system is useful. The pulse width determines the shortest usable gate width thus limiting the range resolution and the ability to reject scattered light. Proper wavelength depends on the scattering media, the spectral response of the imager and the eye safety issues. Jitter of the illumination pulses affects the ranging (gating) resolution of the system.

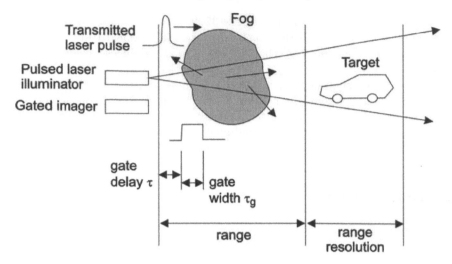

Fig. 5.11. Gated imaging

Short pulses, below 100 ps, are especially suitable for the range-gated imagers operating in photon counting mode. Such imagers based on Geiger mode APD arrays are under development as reviewed in Sect. 5.2 [94]. Besides sensitive gated imaging, Geiger mode photon counting arrays would facilitate direct range imaging as a result of accurate timing properties of the Geiger mode avalanche breakdown. The development of small-sized high power picosecond laser sources would significantly press forward the development of high performance gated imaging systems. Recently a miniature laser system, producing 35 μJ, 1,537 μm pulses of 190 ps (FWHM) duration at 6 kHz repetition rate, was constructed and characterized [98].

5.4 Light Beam Position Measurement Using Position Sensitive Detector (PSD)

Analogue PSD is an optical beam position sensor utilizing, e.g., photodiode surface resistance of lateral effect photodiode (LEP) providing continuous position data with high resolution and high-speed response. LEP can be divided into one-dimensional and two-dimensional types. Position resolution is the minimum detectable light spot displacement indicated by distance on the photosensitive surface. The numerical value of the resolution can be, e.g., 1: 10,000 (1 μm position resolution and 10 mm length of the detector). The precision in outdoor environment is usually limited by atmospheric turbulence.

The LEP produces two (1D) or four (2D) currents which are proportionate to the relative position of the incident light spot. The spot position is calculated from the difference in magnitude of the currents as seen in Fig. 5.12.

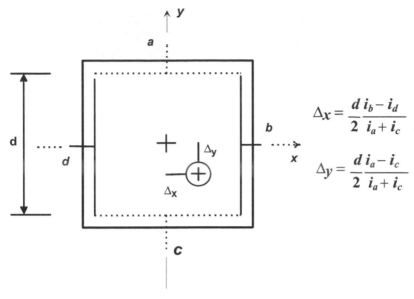

$$\Delta x = \frac{d}{2}\frac{i_b - i_d}{i_a + i_c}$$

$$\Delta y = \frac{d}{2}\frac{i_a - i_c}{i_a + i_c},$$

Fig. 5.12. Operation principle of two-dimensional LEP

When a spot of light falls on the detector's surface, the current from each anode is proportional to the relative position of the spot. If the spot is centered, the currents will be equal. As the spot moves, the output currents change allowing the displacement to be calculated by the formulae given in Fig. 5.12, where i_a, i_b, i_c, and i_d are the average currents of the contacts a, b, c, and d. The length of the side of detector active area is d. The coordinates supplied by the LEP are the center of mass of the illuminated beam. LEP provides a linear transfer function and can be used as a linear displacement or angle sensor of a light beam.

For instance in case of reflected beam measurement shown in Figs. 5.2 and 5.3, the lateral displacement of a reflector $(\Delta X, \Delta Y)$ from the center of the measurement field is [99]

$$\Delta X = (d/2)(L/f_e)[(i_b - i_d)/(i_b + i_d)] \text{ and } \Delta Y = (d/2)(L/f_e)[(i_a - i_c)/(i_a + i_c)],$$

$$(5.11)$$

where L is the distance to the target and f_e is the effective focal length of the receiver optics. Important performance parameters are position resolution, linearity, and the measurement field-of-view (MFOV), which should be optimized in calculating the background radiation.

$$\text{MFOV} = d/f_e. \qquad (5.12)$$

Commercially available LEPs have usually square active area with 1–10 mm side length and are mostly silicon based with spectral response of 400–1,100 nm.

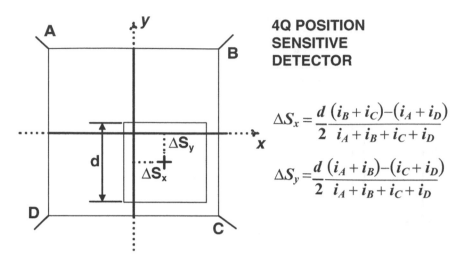

Fig. 5.13. Four quadrant (4Q) position sensitive detector operation principle

Another analogue PSD type is so-called four quadrant detector (4Q) consisting of four separate photodiodes filling each of the quadrants depicted in Fig. 5.13.

4Q PSD gives also four currents to determine the beam position. Disadvantage is in its incompetent linearity and in the condition that the beam must partly cover each of the diodes.

Analogue PSDs favourite in high speed measurements and in the presence of high background radiation. PSDs simple design and operational principle gives the advantages of stability and reliability. The electronics needed for signal processing of the analogue output is quite simple and can be implemented at low cost. Position accuracy of about 0.1% is achievable and the dynamic amplitude range is over several decades. To avoid, for example, interference from stray light, standard method for signal processing such as using modulated light can be easily used.

Alternatives for analogue PSDs are array detectors (CCD and CMOS). Both analogue PSDs and array detectors have the ability to detect light but they do it in a different way. The PSD gives an output that is a function of the centre of gravity of the total light quantity distribution on the active area. The output of the array detector is the peak value of the light intensity over the active area for each pixel and thus describes a picture. The intensity of pixels of CCD is red sequentially, row by row. CMOS array gives the intensity of each pixel in parallel mode.

A CCD is an array of closely spaced MOS diodes. The light is recorded as an electric charge in each diode. The accumulated charges can be transferred in series mode to the output of the device controlled by sequence of clock pulses. The CCD gives a digital output. The CCD cannot measure the centre of gravity of a light spot without additional digital signal processing. Sampling and digitally processing make it slower than the PSD. On the other hand all the pixels have a mask defined positions with high accuracy. However, in order to reach maximum accuracy and the highest resolution the interpolation between neighbouring pixels must take place. The distance between two adjacent pixels sets a lower limit for the spot size. The dynamic range of a CCD is limited and sudden shift in light intensity can give rise to blooming. CMOS arrays overcome many of the CCDs weaknesses when it comes to dynamic range and speed.

A weakness of the analogue PSD is that it cannot differ between many beams. The output is the resulting center of gravity from the whole illumination. Using an array gives the possibility to differ between 2 different beams by evaluating the signal strength in the light spots. As can be seen from the above the fastest way to measure the position of the center of a light spot is to use an analogue PSD. It can be used effectively in alignment systems where the position of a reference laser beam is measured. Such systems can be used for many alignment applications, from bridges to optical shooting training systems.

PSDs are widely used in displacement sensors using triangulation. Such a system can be made at a low cost using rather simple electronics. The disadvantage is that the condition where the light penetrates into porous object like paper, the surface being measured can cause considerable variations in measurement values. Also the texture of the surface may distort the shape of the light spot used for measurements. This can shift the measured centre of gravity of the light spot thus fooling the PSD. Using sophisticated signal processing such as optical filtering and synchronous detection together with a PSD may solve some difficult measurement tasks such as measuring the displacement of hot iron or taking measurements inside the arc of a welding torch.

Line scan camera is an image capturing device having a CCD sensor which is formed by a single line of photosensitive elements (pixels). Therefore, unlike area sensors which generate frames, in this case the image acquisition is made line by line. One scanning line can be considered as a one-dimensional mapping of the brightness of an observed line in grey levels (e.g., from 0 to 255 levels). A sudden change of the grey level in a single point corresponds either to a point on the edge of an object or to any aspect variation of the acquired image. Detection of this change allows a precision measurement, thanks also to the high resolution of the linear sensor which is considerably better than the resolution of an area imager. For instance, by using a backlight, the position and width of a strip or the impurities (holes, scratches, spots, etc.) on the inspected surface can be detected.

Line sensing system has high speed and high resolution. Resolution of line scan camera is approximately 10 times higher than area cameras and it can scan typically at 20 MHz (0.26 ms/5,120 pixel line) compared with 10 MHz (26.2 ms/512 × 512 picture) of area camera. In inspection of the continuously moving object like sheet, the signal processing is easily done by line camera because of its video output by each scan. The area camera needs proper synchronization.

Concerning the applications (Sects. 5.5.5 and 5.5.6) of light beam position measurement the focus of this book is on the usage of LEP.

5.4.1 Resolution and Turbulence

In practical applications PSD can be used in two different ways to measure the beam position or direction. Direct beam sensor use a narrow beam hitting the active area of the PSD and the reflected beam sensor use wide angle beam to illuminate the target and a reflector on it. The reflector can be a corner cube prism or made from reflecting foil. Practically the light reflected from the reflector dominates the signal coming back to the receiver. A positive lens before the detector focuses the light on the surface of the PSD. The position of the focus point gives the light beam direction compared to the reflector. If the reflector is in the middle of the beam the reflected light is focused on the middle of the PSD active surface. If the beam is scanned, the light spot on the PSD is scanning similarly. The distance scales the actual movement of the light spot on the PSD. The direct beam sensing was depicted in Fig. 5.1, and the reflected beam sensing in Fig. 5.2.

The resolution of the LEP detector depends on the signal amplitude. The accuracy of signal detection has been analyzed by several authors, recently by Donati et al. [100]. Signal based shot noise (quantum limit) sets the ultimate limit for the resolution. It is inversely proportional to the number of detected photons or more accurately to the square root of the number of photoelectrons (N_{ph}) used in the measurement process ($\sim 1/\sqrt{N_{ph}}$). In real case usually the detector dark current and preamplifier noises are dominant with respect to signal based quantum noise. The other factor is so called characteristic length (L_c) which depicts distance, diameter or time needed in signal processing. The longer the characteristic length the bigger the noise (σ),

$$\sigma = L_c/N_{ph}. \tag{5.13}$$

The characteristic length for analogue PSD is the diameter (w) of the light spot on the detector surface. It can be used to calculate the position measurement precision ($\sigma_x,$ σ_y), as seen in (5.14).

$$\sigma_x, \sigma_y = (\pi/2)^{1/2}(kw/SNR), \tag{5.14}$$

where k is a constant determined by PSD noise parameters and SNR is the ratio between the root mean square (rms) signal and noise currents [100]. If the reflected beam method is used the beam width on the detector surface is determined by MFOV, as depicted in (5.12). The characteristic length for different measurement systems is determined by the measurement principle, i.e., for pulsed TOF range finder it is the time needed to make the time pick off and for triangulation the uncertainty in angle measurement [100].

The atmospheric turbulence has been analyzed and measured by Mäkynen et al. [99, 101, 102] and recently in [103]. Atmospheric turbulence causes phenomena commonly referred to as beam wander, scintillation and beam breathing, according to the effect produced on the beam spot as seen on the screen after travelling through a turbulent atmosphere. Beam wander means random changes in the position of the beam spot on the screen, scintillation illumination fluctuations within the beam and breathing expansion and contraction on the spot beyond the dimension predicted by its geometry and diffraction. These effects are caused by index-of-refraction inhomogeneities in air which mainly arise from spatial temperature differences within the atmosphere. The dominant effect is beam wander in weak turbulence and scintillation in strong turbulence.

The effect of atmospheric turbulence on beam propagation is a result of complicated phenomena. In certain cases, however, it can satisfactorily be estimated using simple equations. Geometric optical formulation can be successfully used to predict large-scale effects such as angle-of-arrival fluctuations or beam wander. This presupposes, however, that the beam diameter (w) should not be markedly altered by diffraction, or by scattering due to atmospheric turbulence. The first condition is fulfilled if the measurement distance (L) is within the Fresnel diffraction range ($w > \sqrt{L\lambda}$) and the second condition if the path-integrated turbulence is weak (coherence length $\rho_o > L\lambda/w$) [104].

There are many outdoor applications in which optical measurements are performed under strong atmospheric turbulence. These include various forms of displacement measurements such as laser-based alignment sensing, surveying, or optical shooting simulation, for example. In these applications the measurement distances vary typically from tens of meters to about 1 km, and the desired accuracy from 0.01 to 0.1 mrad corresponding to a few millimetres of displacement at a 100 m distance, for example. The calculations made in [99–105] show that such accuracy level could be achieved.

The effect of turbulence differs from each other if direct or reflected beam principle is used. Also the PSD principle, LEP or 4Q, may have an effect on the result. Figure 5.14 shows principally the measurement arrangement made by Mäkynen [69, 105] to verify the theoretical calculations.

Shortly, given the measurement results using LEP as a detector show turbulence limited resolution to vary from 0.3 to 4.5 mm as distance increases from 50 to 300 m at intermediate turbulence level and for direct beam measurement. The resolution of the reflected beam method was about 50% worse with corner cubes as reflector. Four quadrant detectors (4Q) gave in outdoor conditions about 10 times worse resolution than LEP. The above conditions are not too restrictive when considering practical reflected beam sensors. Turbulence can deteriorate also the accuracy of pulsed TOF laser rangefinder if not taken carefully into account in designing the device. The received signal amplitude is affected by the scintillation.

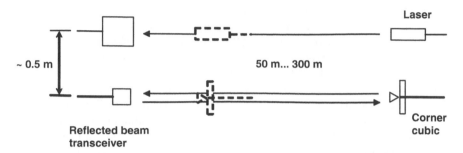

Fig. 5.14. Atmospheric turbulence measurements in direct beam and reflected beam arrangement as a function of distance

5.5 Applications

Measurement of laser pulse transit time and laser beam position has a lot of applications. In the following some recently developed applications for analogue PSD (LEP) and pulsed time-of-flight distance measurement will be discussed. Some of them are under development in research laboratories but most of them have been commercialized.

5.5.1 Traffic Control Applications

Inductive loops have been widely used for years in a variety of traffic control applications. However, modern optoelectronic measurement technology offers a very worthwhile alternative, not only technically but also with a competitive price. There are clear benefits for using optical measurement, too. Not only are they simple to install and maintain, but also new possibilities open up. The earliest well-known application for optoelectronics in the traffic industry was long distance speed radar. The technology has been developing fast since those first units were used and today laser range finding measurements are probably the leading speed checking technology.

Small, integrated laser distance measurement sensors are well suited for many traffic applications, including [106]:

(1) Distance measurement-based camera triggering for license plate recognition (LPR) systems.
(2) Vehicle speed measurement and LPR based camera triggering in speed violation cases.
(3) Vehicle average speed measurement between two sites.
(4) Vehicle classification, profile, height, and length measurement.
(5) Headway or distance between vehicles.
(6) Traffic light control, left turn sensor.
(7) Speed measurement for portable speed cameras.

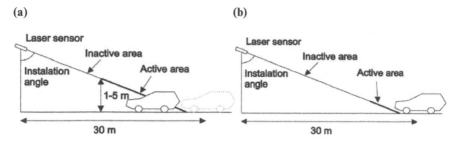

(a)   (b)

Fig. 5.15. (a) Speed measurement and (b) LPR camera triggering

(8) Tunnel entrance control, etc.

The basic distance information can be directly used in some cases, but usually some signal processing is necessary. Often measurements are taken thousands of times a second, demanding a lot of signal processing capacity.

One application for a laser pulse TOF range finder is vehicle detection, when the vehicle is coming toward the LPR camera. This can be done also for departing vehicles. The operation principle is to quickly measure the distance to the object and use that to define the exact triggering moment. Typical sensor installation is 5–7 m above the ground looking forward, down at the road. When the vehicle enters the trigger area, defined by the parameters, the sensor sends a pulse to the camera. In such cases the triggering accuracy has to be as high as 5–10 cm (or one millisecond), depending on the installation, Fig. 5.15b.

Similar installations can be used for measuring the speed of approaching or departing vehicles, Fig. 5.15b. The speed measurement should be very fast and the information should be available 20–30 ms after the vehicle has passed the measurement point. The speed has to be measured in the range of 10–250 km h^{-1}. To give the overspeed trigger information for the camera, measurement can be arranged in two phases. First, the unit calculates a rough speed value, which it uses for the trigger. This can be done in a few milliseconds. Then the signal is processed further to give the final speed value to the camera system. This allows reliable measurement, even if vehicles are driving near each other. The information can be used to control distances between vehicles, too. One of the best features of the optical principle is the possibility to measure exactly the target that is pointed, because the measurement field of view can be adjusted to as small a value as needed. A microwave radar has usually more than 10° measurement beam and the user often cannot be sure which vehicle will be detected if more than are in the field of view.

A fast distance measurement with scanning mechanics (laser radar) is used in industrial applications for profile measurement or area protection. In traffic control, fast distance measurement without scanning mechanics can be used for vehicle profiling, because the vehicle itself is moving. A fast range finder can generate a profile that can then be used for classification purposes.

Combined with speed measurement, the height and length of the vehicle and the profile data can then be used for further analysis of the vehicle type.

An alternative for that kind of application is to use two-line scan cameras on the side of the road. As the vehicle travels across the vertical scanning lines, two images are formed and can be correlated with each other to determine the vehicle speed. In addition to speed, the system is able to extract a broad range of traffic information, among others the vehicle size, acceleration and intervehicle distances. This technique can measure several lanes but heavy traffic may result in wrong alarms [107].

A laser rangefinder is also a good option for intersection control. The devices controlling several lanes can be installed in one place to make installation and handling easier. Each rangefinder may be directed to measure its own lane. They can also be used as a vehicle detector for changing lights or to check for red light violation. Many control system manufacturers and system integrators have seen the benefits of laser rangefinder in their systems. The easy installation, multifunctionality, adaptivity, and maintainability are benefits of optoelectronic sensors based on range finding [106].

A typical laser range finder for traffic applications is a water tight, nitrogen-filled device which allows distance measurement to poorly reflecting surfaces at high speed with a very good resolution. The low power consumption device can be used in both fixed installation and portable systems in varying temperatures and environments. In any measurement where people are involved, it is essential that the laser device fulfils the appropriate international laser safety standards. They define the level of laser power used as well as some other technical limits. The standard splits the products into several classes, starting with class 1. The required level of safety in traffic control is class 1, where the measurement is "eye safe". Because the limitations of class 1 are high, the technology must be optimized to allow effective measurement.

5.5.2 Medical Applications

When a turbid material is illuminated by a short laser pulse it temporally spreads and three types of photons traveling through material appear: ballistic, snake, and diffusive, depicted in Fig. 5.16.

The transit time is shortest for ballistic photon and longest for diffusive photons. The phenomenon is also called photon migration. In each case light travels a different distance through the medium, and its time-of-flight depends on the distance traveled and the refractive index of the medium.

Both nonscattered photons and photons undergoing forward-directed single-step scattering contribute to the intensity of the ballistic component (composed of photons traveling straight along the laser beam). This component is subject to exponential attenuation with increasing sample thickness. The group of snake photons with zigzag trajectories includes photons that have experienced only a few collisions each. They propagate along trajectories

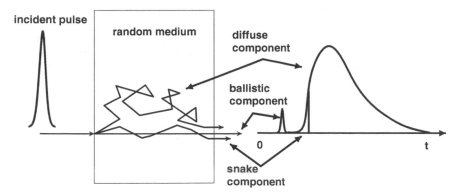

Fig. 5.16. A light pulse propagating through a random medium spreads into ballistic, snake, and diffuse components

that deviate only slightly from the direction of the incident beam and form the first-arriving part of the diffusive component. These photons carry information about the optical properties of the random medium and parameters of any foreign object which they may happen to come across during their progress.

The diffusive component is very broad and intense since it contains the bulk of incident photons after they have participated in many scattering acts and therefore migrate in different directions and have different path lengths. Moreover, the diffusive component carries information about the optical properties of the scattering medium, and its deformation may reflect the presence of local inhomogeneities in the medium [108].

Photon migration can be studied using streak camera. It can measure the pulse shape of the short laser pulse traveling through the scattering object with picoseconds accuracy, Fig. 5.17a. Another method is to use time-of-flight method (pulsed or phase shift) using a technique similar to that used in distance measurement, as depicted in Fig. 5.17b.

Streak camera can measure the path length distribution and find the range of the average path length of photons. If reflected photons are detected as a function of time (gated imaging, see Sect. 5.3.), imaging through turbid media can be achieved.

Time-resolved transmission in medical application is used, for example, for breast cancer detection. The measurement principle is similar to that used in laser radar in Sect. 5.2. Light sources and detectors are usually fiber pigtailed. Light sources with different wavelengths can be used to optimize the measurement. The technique is based on unequal optical properties of cancer tissue compared to normal tissue. Variations in optical properties of normal breast tissue set limits to the performance of such techniques [109, 110].

First proposed in the late 1980s, photon migration techniques for detection of breast cancer have since been steadily increasing in terms of rate of devel-

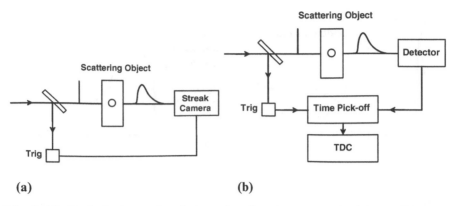

Fig. 5.17. Typical schemes for photon migration measurements, (**a**) streak camera, and (**b**) range finder

opment. Like X-ray mammography, optical mammography is a noninvasive technique, but without the potential risks involved with the use of ionizing radiation. It is clear, however, that the spatial resolution of optical images is inferior to that obtained by X-rays due to the strong scattering at optical wavelengths. Instead, the great potential for optical mammography lies in functional imaging, providing physiological information by spectrometric quantification of parameters such as tissue constituents and oxygenation. The used wavelength is normally in the near infrared (NIR), where tissue absorption is relatively low and transmittance measurements are therefore feasible. Several optical mammography instruments have already been constructed [111].

Optical coherence tomography (OCT), as an interferometer method, is intensively studied for medical applications, too. The first applications were related to the evaluation of biological tissue showing the cross-section of retinal structures. Skin investigations are increasing since usage of clearing agents to increase the penetration depth [112].

5.5.3 Industrial Applications

In addition to optical properties of paper the optical methods can be used in paper manufacturing processes to study pulp consistency, fines content and filler content, and the thickness, basis weight, density, and porosity of paper. Only the investigations to test paper properties by photon migration are introduced here. As paper is a complex material most properties are measured with a standardized test, which is related to, but is not necessarily an accurate measure of the desired property. For example, airflow measurements are used for measuring paper surface roughness.

One approach to optical paper inspection is time-resolved spectroscopy. To be able to use it the basic knowledge of photon migration in paper has to be known. Can paper thickness, basis weight and structure be measured?

Can they be done by studying photon migration? E.g., the character of paper porosity is complex. If porosity is considered to be the relative volume of pores or of air the following formula may be used:

$$\varphi = (V - V_f)/V = 1 - \rho/\rho_f, \tag{5.15}$$

where φ, porosity; V, volume of paper; V_f, volume of fibers; ρ, density of paper; ρ_f, density of fibers $= 1,500 \, \text{kg m}^{-3}$ for perfect cellulose fibrils. Again as the paper surface is not clear these definitions are somewhat inaccurate. People interested in printability have their own definition and measure of porosity. If air is blown through paper and the flow is measured it corresponds to the amount and size of the pores through paper and thus gives an estimate of how ink penetrates to the pores [113].

Optically, paper consists of a large number of closely packed single particles, whose shape and refractive index varies depending on whether they are fibers, fines, or fillers. Since fines are fractions of fibers, they have the same refractive index as the part of the fiber they are made of. However, fines have relatively larger scattering cross-section than unbroken fibers.

Carlsson et al. [88] *described* a method for the time-resolved recording of light scattering with a streak camera in thin highly scattering media. The method was applied to paper. Then they studied the dependence of light scattering on basis weight and density.

Time-of-flight method can be used in a similar way as described in Sect. 5.5.2 for paper investigation. The porosity (φ), density (ρ), and thickness of paper can be adjusted by pressing the paper sheet. Figure 5.18 shows photon migration measurement results in compressed paper made by Saarela [113].

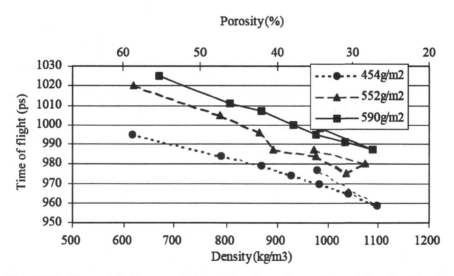

Fig. 5.18. Time-of-flight presented as a function of porosity and density. The delay was measured with a modified range finder. The laser had a wavelength of 650 nm [113]

The paper samples were pressed between two hardened glass plates to adjust the thickness during a measurement session to change the density. Furthermore, if porosity is defined as the volume of air in paper, the results show that the pulsed laser TOF measurement can be used as a measure of paper porosity. A detailed description of the experiment can be found in [114].

The results show that pressing paper leads to a corresponding decrease in TOF. This indicates that the distances between the scattering cross sections of fibers decreases. However, if the basis weight is lower than $200\,\mathrm{g\,m^{-2}}$, pressure has little effect on time-of-flight. Furthermore, the results show that the TOF of a laser pulse can be used as a measure of porosity in paper. However, this necessitates a study on the effects of fillers and fines.

Karppinen et al. [115] performed a related study on papermaking pulp properties using time-of-flight (TOF) measurements. The pulp samples they tested were of low consistency, the maximum value being 0.6%. They concluded that the TOF measurement technique is best suited for measuring fines content.

Laser radar has many other industrial applications. One demanding application is seen in Fig. 5.9a. This application is concerned with checking the thickness of hot refractory linings in steel works. The temperature of the lining during measurement is high, usually between 1,100 and 1,400 °C, which induces severe background radiation and noise problems. These measurements show that it is possible to use the pulsed laser radar technique in demanding measurement applications of this kind to obtain reliable data on the lining wear rate of a hot converter in a steel works [85].

OCT can be used also to study tomographically the structure of paper [76]. Paper thickness is typically $100\,\mu\mathrm{m}$. OCT can be used only near the surface. Using clearing liquids the measurement depth can be increased. In addition to structure measurements the technique can be used to study liquid penetration into paper [116]. OCT technique was developed to human tissue measurements. However, it has recently found many industrial applications [117].

Optical fiber sensors have many functioning principles and applications. They have strengths, such as ruggedness to high temperatures, harsh environments, and electromagnetic interference. Fiber optic sensors are also long-lived, compact and flexible making them attractive for embedding into organic (the human body) and inorganic structures (bridges, buildings, and vehicles). The multiplexing capability of optical sensors makes them an excellent candidate for sensor networks.

A Bragg grating can be embedded in a fiber and used to measure strain based on length variation of the grating. It measures point-like strain. Long optical fiber can be used as a strain sensor by measuring its length variations using TOF principle. Integral strain and point strain are both important in many applications. Fiber optic sensor can be embedded during manufacturing process in a composite material to determine the strain state of the structure. Examples of such structures are fiber-reinforced composite materials which are increasingly used as engineering materials in aircraft, buildings, containers, paper machines, and in shipbuilding. Composite materials are often designed to carry heavy loading and they are used in many critical structures. Optical fiber sensors embedded in composite provide a good opportunity to control the behaviour of these structures. Optical fiber sensors are ideal solution for measure strains and temperature inside composite structure, because their material is often identical to reinforced fibers and because their effect on material strength is negligible due to their small size [118]. An example of such structures is a rotating composite cylinder of paper machine. It is demanded to monitor the condition of the cylinder during the paper manufacturing process. A wireless strain monitoring system using optical fiber sensors embedded in a composite material during its manufacturing process is presented in [119]. However, composite cylinders posed severe mounting problems.

Point strain can be used in monitoring certain control points, and integral strain can be used to monitor the whole structure. The length of a standard Bragg grating is typically 10 mm. The performance of a structure with an inhomogeneous strain distribution can be more reliably measured by a long-gauge sensor. A combination of the FGB (Fiber Bragg Grating) interrogation system and the fiber-optic integral TOF strain analyzer make good use of the advantages of both systems. Like in the case of focal plane laser radar, the TDC (Time to Digital Converter) can be developed for multiple stop pulses, e.g., reflected from in-fiber FBGs. This is the way to use one optical fiber as a sensor network for both the integral and point like strain measurements [120].

5.5.4 Monitoring of Bridges

Fiber optic sensor based on TOF fiber length measurement applies well to the study of such large structures as bridges and dams, where both dynamic and static strains and their derivatives such as cracks, deflections, and displacements are to be measured at many locations.

The static and a dynamic strain measurement of a bridge are important for monitoring health and for repairing schedules. The usage of time-of-flight fiber optical strain sensor has been studied for strain measurement affected by the traffic for a bridge [121]. The sensor was installed in the bridge near the center of the river as shown in Fig. 5.19.

When a lorry passes over the bridge, the bridge vibrates up and down. The structure between the two fixed points for the sensing area alternately endures compression and tension. One test result is shown in Fig. 5.20. The peak value of the strain is about 30 μ strain.

Filtering the measurement data decreases the noise amplitude, but it slows down the measurement speed. The system achieves a precision below 1 μstrain for sensor lengths up to 10 m and a measurement frequency of 100 Hz. With the measured strain and the physical parameters of the sensor and the bridge, the vibration amplitude of the bridge can be calculated. This information is useful in bridge performance evaluation, mathematical analytical model building and fatigue analysis.

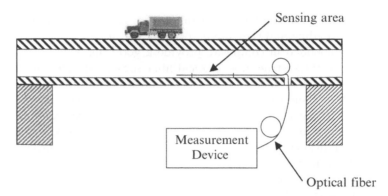

Sensing area

Measurement
Device

Optical fiber

Fig. 5.19. The location of the installed optical fiber strain sensor for bridge deck strain monitoring. The length of the sensing area is 2 m (Courtesy of Mr. G. Duan)

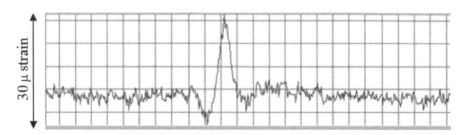

30 μ strain

Fig. 5.20. Integral strain measurement of a bridge when a lorry is passing the bridge (G. Duan)

PSD and laser beam is an alternative method in measurement of the dynamics of large constructions such as bridges, towers, buildings and masts, and movement of other moving objects. The laser diode transmitter and receiver are installed so that the laser beam forms a reference line from the transmitter to the center of the receiving target area. When the receiver moves due to the motion of the part to which it is connected with respect to the reference laser line, the position of the laser beam on the optical target changes correspondingly. The electronics inside the receiver measures the centroid position of the laser beam on the optical target continuously and supply data on its x and y coordinates continuously to a PC computer, for example as seen in Fig. 5.21.

Figure 5.22a shows Kemijoki bridge in Rovaniemi, Finland. Using previous measurement principle the pylon top bending in both directions, main deck deflection in the middle section and main deck torsion were measured. The measurements were performed with one receiver at the top of the pylon and two on the deck. The transmitters were installed on the fixed foot of the pylon to avoid beam movement during the measurement. An example of the deck deflection is seen in Fig. 5.22b.

(a)   (b)

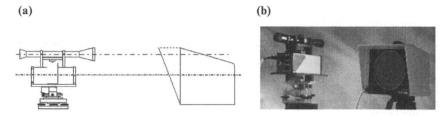

Fig. 5.21. (a) A schematic and (b) a photograph of laser head and PSD system for measuring the movement and alignment of a structure (Noptel Oy)

(a)   (b)

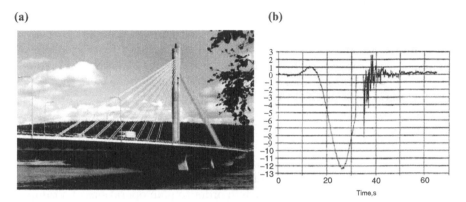

Fig. 5.22. (a) Kemijoki bridge and (b) its deck displacement (mm) when a truck is passing the measurement position

Loading was performed using trucks filled with sand to a gross weight of 30 tons, the purpose was to compare the design parameters and values with those measured under real conditions. Compared to fiber optic strain measurement here the displacement of the structure is displayed.

5.5.5 Railway Track Measuring and Guidance of the Tamping Machine

A PSD can be used to obtain displacement reading of the measured object relative to the laser beam with high accuracy. A narrow and low divergent laser beam forms a straight line between two points. This laser beam can be used to align or level assembly tracks, conveyor lines, etc. The PSD is fixed on the object and the laser beam hits the PSDs active area, as depicted in Fig. 5.21. The active area can be enlarged using PSD diffusive screen and Fresnel lens. For example, sensing of 200 mm can be achieved with 10 mm PSD diameter. However, this decreases the resolution 20 times worse compared to PSD resolution.

The measurement can be implemented also using reflecting geometry, Fig. 5.23. A reflector is fastened to the object. Transmitter and receiver are

(a)   **(b)**

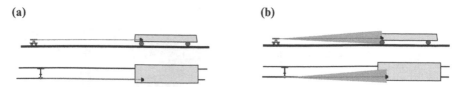

Fig. 5.23. Guidance of a tamping machine. (a) Narrow direct beam, (b) wide angle reflecting beam

in the same place. Because the light travels twice the measuring range, the resolution worsens by about 50%, see Sect. 5.4.1.

One application of the PSD method is a measurement system for railway track machinery. It can be used with tamping machines for measuring the alignment of rails in accordance with specific guidelines [122]. Similar system is also used with coaches, cranes and lightweight trolleys to measure the vertical and horizontal position of the rails.

The system employs the laser beam position measurement device to access the position of the rail in the horizontal and vertical planes using a receiver connected to a part of the vehicle, which is in direct contact with the rail. The system can be installed to many types of machinery and does not require additional automation. The system can be used during the rehabilitation operation or to measure the rail position in order to plan operations of a tamping machine in advance or inspect the result of the work afterwards.

A PC allows the operator to monitor the straightness of the line rail graphically in both directions, set values for lifting, alignment and levelling, and store measurement data in a numerical file as required. The operator directs a safe, visible laser beam at the receiver and locks it in a steady position. The receiver recognises the beam on the optical target and measures its position accurately in milliseconds, Fig. 5.23.

In both measurement principles the laser is installed on a trolley. When using a narrow beam the receiver is fastened on a tamping machine and using a wide beam the receiver is set on the laser trolley and a reflector is fastened on the tamping machine. In the reflecting measurement case the distance is needed for scaling the result. The laser trolley is positioned at fixed point and fastened to the reference rail. Then the laser is adjusted to hit the receiver or the reflector on the tamping machine. Good environmental durability can be achieved by proper modulation of the laser beam, optics design and the detection technology employed in the receiver to minimize the influence of sunlight, temperature, fog, rain, snow, etc. on the accuracy over a wide measurement range. The transmitter can be at a distance up to several hundreds of metres from the receiver, depending on the environmental conditions.

This kind of system can measure the alignment of the railway track horizontally and vertically in good weather conditions with mm-level accuracy. One measurement round can be typically 500 m. Wide angle reflecting beam system is simpler to install, but the reading must be scaled by distance. This kind of measurement helps the track reinforcement work to adjust the rail rapidly and precisely to reference position. The correction values for level and alignment are calculated from the difference between the nominal and actual position. The front chord end of the tamping machine is then automatically guided and controlled according to these correction values.

5.5.6 Marksmanship Training

A reflected beam PSD method can be used for shooting training, too. Marksman aims his gun to the target and fires. The only information left is a hole on the target. In training it is important to follow the aiming before and after pulling the trigger. Using a reflector on the target, light source to illuminate the target, a PSD equipped with lens at the same axis as the transmitter, the device orientation against the target can be recorded, Fig. 5.24a. The device is fastened on the sport weapon and its orientation path on the target surface can be determined before and after the shot. Every shot leaves a unique track, which provides information on how the shot was generated, Fig. 5.24b.

The accuracy of 0.1 ring (standardized distance and sport shooting target) can be achieved at indoor ranges and in good weather conditions also at outdoor ranges. The degree of difficulty can be scaled electronically for shorter distances to approximate the normal distance. This enables a thorough analysis of the shooter's training event.

The development of the shooting skill follows the principles generally valid for the development of any skill. As with any other skill, that of shooting has to be built up gradually from first principles, moving from simple to more complicated tasks. A trainee needs [123]:

(a)   **(b)**

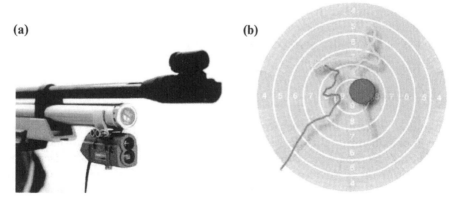

Fig. 5.24. (a) Measurement device of the shooting training system. (b) Gun orientation path on the target surface before and after the triggering (Courtesy of Noptel Oy)

(1) To concentrate on one thing at a time when learning a complex skill.
(2) To perform enough successful repetitions to ensure that performance becomes as automatic as possible.
(3) Direct and immediate feedback, preferably in real time, will make learning more efficient.
(4) Elimination of external interference factors as far as possible when developing individual aspects of the skill.
(5) Motivation for practising, as a lack of motivation can frustrate even the most sophisticated training system.

Advanced measuring techniques allow feedback in classroom and range conditions. The training equipment allows using the blanks and pneumatic recoil system in the trainee's own weapon. Training safety is increased, environmental burden is decreased and money is saved.

The shooter receives immediate, objective feedback on the shot from the PC screen, so as to make the necessary corrections before the next shot or series, thus making the training very efficient. In fact, the sooner the shooter can correct any bad shooting habits, the sooner he/she will learn the right shooting technique. Faulty techniques can mean a lot of extra work and unnecessary time spent on the way to becoming a skilled shooter. This kind of optical aid enables a thorough analysis of the shooter's skill to be made.

Augmented information feedback has been considered important for learning motor skills. It means supplementing information, which is provided as extrinsic feedback to a learner. The mode of feedback has been shown to have an effect on motor skill learning, especially in studies performed in laboratory settings. Applications of motor learning theories to teaching of real-word tasks, e.g., of shooting sport skills, are the final goal of motor learning studies. The study reported in [124] showed that the feedback on shooting scores improved significantly the sum of scores among novice shooters during 12-weeks training period. It shoved also that there may be several uncontrollable variables which may interfere with the motor learning process and put special demands on the studies in regard to real-word tasks.

5.6 Conclusions

There is a demand for industrial measurement techniques and the demands on manufacturing quality increase simultaneously. Traffic control has many applications for noncontact measurement techniques. Many sport disciplines need measurement devices to get feedback on training to develop physical and mental capacity. Medical diagnostic measurements need also new faster and more reliable sensors. There is a wider interest to monitor large structures like bridges, buildings and dams to prevent accidents caused by their failures. Optical sensors may give new possibilities for the measurement needs previewed.

Optical distance and beam position measurement plays an important role as it has many advantages in comparison to conventional methods: contact-free, quick measurements at high resolution and usable in harsh environments (EMI, temperature, etc.). However, the methods are based on several measurement principles (time-of-flight with many implementations, triangulation, interferometry, fiber sensors, imaging processes, alignment, auto focus, etc.) Every principle has its specific properties and thus specific advantages and demands for different applications.

For instance, triangulation has become a standard method for short range distance and shape measurements. Triangulation is not generally suitable for measurement of edges or holes due to shadowing effects. Time-of-flight distance measurement offers solutions for longer distances and without shadowing effect. Pulsed TOF is well suited for moving noncooperative targets, because the single pulse touches the object very shortly. Pulse duration is typically nanoseconds. Phase shift method has good accuracy when measuring to cooperative nonmoving target (reflector).

For instance a fast growing application area is at the moment traffic control, where many automatic control systems need reliable measurement in difficult environment. TOF sensors are in use as intelligent traffic camera triggering, vehicle profile measurement and vehicle classification as well as in speed measurement. The appropriate measurement method must always be identified, optimized and customized for the specific application.

6

Laser Velocimetry

Laser velocimetry, often referred to as laser anemometry and laser flowmetry and also known by some other names which will be used below, is a set of laser techniques designed for noncontact, remote measurements of velocities of gaseous, fluid, and solid media. Basically all these techniques make use of light scattering by tracer particles naturally present in the gaseous, fluid, and tissue-like media or specially administered therein, or light scattering inhomogeneities and roughness on the solid surfaces. In this sense, laser velocimetry is based on light scattering theory, in most cases within the Mie theory approximation. There exist also laser velocimetry techniques based on detecting laser-induced fluorescence signal from moving fluorescing species. However, this set of techniques is beyond the scope of this chapter.

Being able to perform nonperturbing velocity measurements is very important for solving various basic problems of experimental fluid mechanics and gas dynamics, as well as of biomechanics, in particular, hemodynamics. The applications of different variants of laser velocimetry range from the study of aircraft and vehicles, combustion, multiphase flows, channel and cavity flows, boundary layers, natural convection, unsteady flows and flow instability, vortices, turbulence, and large-scale environmental flows, to low-Reynolds-number and microscale flows, in particular, in medical/physiological applications. In all these applications, the noncontact, noninvasive, and nondestructive mode of measurement is of utmost importance.

The intrinsic limitation of laser velocimetry techniques is the requirement of relative transparency of the flowing medium and the medium surrounding the embedded flow so that laser light can penetrate therein and one can detect a considerable amount of scattered light. Techniques that allow optical velocity measurements in media with strong scattering are now being developed and will be discussed in this chapter. However, strong scattering implies a drastic limitation in the accessible depth of measurement. Alternative techniques that allow velocity measurements at larger depths are based on using ultrasonic waves and the optoacoustic effect. In some applications, the information on the velocities inside opaque media can be obtained noninvasively by measuring the electric impedance and by other nonoptical techniques.

In general, velocity measurements are made in the Lagrangian or Eulerian frames of reference. Lagrangian methods assign a velocity to a tracer particle at a given time, whereas Eulerian methods assign a velocity to a volume of the measurement domain at a given time.

6.1 Laser Doppler Velocimetry

Laser Doppler velocimetry (LDV) is a technique that allows the measurement of velocity at a local volume in a flow field with a high temporal resolution. When a micron-sized liquid or solid particle or a gas bubble entrains in a gaseous or fluid flow and passes through a laser beam or the intersection of two or more laser beams, the scattered light detected from the particle fluctuates in intensity. LDV makes use of the fact that the frequency of this fluctuation is due to the Doppler shift between the incident and scattered light, and therefore is proportional to the projection of the particle velocity onto the so-called sensitivity vector of the system defined by the relative positioning of the illuminating and receiving optical elements. After the pioneering work of Yeh and Cummins [125], this technique has experienced fast development, allowing now a variety of applications. Some of them will be briefly described in this chapter. Detailed descriptions on the LDV technique can be found in many handbooks [126], article collections [127], and conference proceedings [128,129]. Special arrangements, such as phase Doppler interferometry, allow particle sizing to be performed simultaneously with their velocity measurements. The measurement of particle size distribution and their mean values is based on the laser light wavelength, which is known to high accuracy and is independent of the light intensity.

In different practical applications, depending on the particular aims and class of problems to be solved, single, dual, and multiple beam arrangements of laser Doppler velocimeters are used. In the single beam arrangement, also called monostatic, the studied object or medium is probed just with one beam. The receiving unit detects either only the Doppler-shifted light scattered by moving particles or both Doppler-shifted and unshifted light in the event when some stationary elements also scatter the probing beam in the direction of the receiver. The first case is typical, e.g., of measurements from atmospheric aerosols and nontransparent reflecting surfaces. The second case is typical of physiological measurements, in particular, of blood perfusion in tissue where light is scattered by both moving red blood cells and stationary skin cells, or of cytoplasmic streaming in living cells where the probing beam is scattered by both moving intracellular organelles and the cell's stationary cortex. In this case, the measurements are typically performed without sensing the flow direction. In the single beam arrangement, the detection volume is defined by focusing the probing beam and by the depth of focus of the receiving unit or by the pulse duration in the case of pulsed light sources. Such velocimeters allow measurements of only the online velocity component of the flow. A special case of single-beam velocity sensors is based on so-called self-mixing detection of the scattered light.

In order to measure two components of the velocity and distinguish the direction of the flow, double-beam arrangements, also called bi-static, are typically used. In this case, the flow is probed by two coherent beams, emerging most often from the same laser source. The beams intersect at a distance from the instrument inside the studied flow to form a cross-beam fringe pattern in the intersection (detection) volume. The characteristic lateral size of this volume can be made much smaller than in monostatic arrangements, ranging from a few micrometers in laser Doppler microscopes and typically several hundred micrometers to several millimeters in most of the industrial applications requiring long-range measurements. The size of the detection volume can be reduced by employing a spatial mask in a side-scatter detection, which necessitates two optical accesses to the flow. A higher spatial resolution can be achieved by reducing the size of the detection volume by strongly focusing the laser beams. However, this may cause nonuniformity of the fringe spacing inside the detection volume and smaller numbers of fringes, which will lower the accuracy of velocity measurement and induce apparent turbulence intensity [130].

Shifting the light frequency in one of the beams, e.g., with a conventional electro-optic or acousto-optic modulator, results in a moving fringe pattern in the detection volume. Because the direction of fringe movement depends only on the geometry of the velocimeter and does not depend on flow direction, the velocimeter becomes sensitive to the sign of the flow. However, instead of using frequency-shift elements, which are bulky and difficult to align during assembly, directional discrimination of the fluid flow can be achieved by using a quadrature homodyne technique based on the employment of two laser wavelengths, which generate two interference fringe systems with a phase shift of a quarter of the common fringe spacing [131]. Measurement signal pairs with a direction-dependent phase shift of $\pm\pi/2$ are generated. Signal processing is performed by the cross-correlation technique. The setup provides a constant phase shift of $\pi/2$ throughout the entire detection volume with both single-mode and multimode radiation. The directional discrimination was successfully verified with wind tunnel measurements. The technique offers the potential of building miniaturized measurement heads that can be integrated, e.g., into wind tunnel models.

In order to measure all three components of the velocity vector, three or multiple-beam arrangements of velocimeters are used. There are arguments for using four beams as the minimum requirement for complete three-dimensional (3D) velocity reconstruction, even though three beams supply the three velocity components [132]. To ascertain the velocity components, it is possible also to scan the specimen in a precise manner relative to the point of focus of the beams instead of shifting frequency of the probing beams. The results obtained with these two methods are equivalent. However, scanning is mechanically simpler than frequency shifting and also allows the formation of velocity images – images of the flow velocity over a region in 2D or 3D space.

Under certain conditions, the LDV technique can be used for measuring the velocity gradients, which is of great interest in fluid mechanics. Detailed information about the velocity distribution is a prerequisite for the defined design of aerodynamic devices such as aircraft wings, micronozzles, etc. Another typical application in tube flows is the measurement of velocity profiles, from which the flow rate can be evaluated exactly. In the field of medicine there is special interest in the local resolution of velocity fields in blood veins and arteries of different sizes.

Most commonly, a quasi-pointlike measurement technique is used for this purpose. This means that the velocity information on the whole fluid field is registered not simultaneously but only by local sectioning. For the measurement of the whole velocity field, mechanical scanning is afforded. The spatial resolution is determined by the finite size of the detection volume. Since its length in conventional systems is about 1 mm, the resolution of strong velocity gradients is difficult. To overcome this drawback, the following two methods can be used:

(a) Spatial restriction through the receiving optics, i.e., reduction of the acceptance field of the detector; and
(b) Spatial restriction through the illuminating optics, i.e., reduction of the length of the detection volume.

In method (a), the restriction of the acceptance field of the detector is usually achieved by means of beam stops or confocal imaging. As a consequence, much effort has to be made to the adjustment of the detection unit and only a part of the laser power is used. In method (b), strong focusing is basically used, as mentioned above. However, this usually implies limitation to a short working distance. Such applications are typical of biomedical studies. For example, in [133] the migration of platelets and red blood cells in highly dilute suspensions was studied. For this purpose the detection volume with a lateral size of 5.7 μm and a length of 19 μm was provided and the velocity profile in a rectangular flow channel of width 100 μm was measured. Such a short measurement volume was obtained by using a small focus, resulting in a working distance of less than 4 mm and a crossing half-angle of 17.2°. Other examples of the so-called laser Doppler microscopes and their applications will be discussed below.

In many applications, longer working distances are required. Since the entrance window to a flow tube is of a finite size, the beam-crossing angle is limited for a given working distance and the size of the detection volume cannot be below a certain value. Furthermore, the variation of the fringe spacing increases inversely proportional to the square of the diameter of the beam waist in the detection volume. Since the longitudinal position of the particle passing through the detection volume is usually not known, the spectral width of the Doppler line broadens, which can be regarded as virtual turbulence. Therefore, the design of a LDV system is a compromise between high spatial resolution (given by the length of the detection volume) and the accuracy of the velocity measurement (given by the variation of the fringe spacing).

The advantages of fiber-optic transmission and receiving LDV systems, such as flexibility and immunity against electromagnetic disturbances, are well recognized. Single-mode fibers are more frequently employed in conventional devices. The degradation in quality of the signal has prevented the employment of multimode fibers for beam delivery for a long time. However, multimode fibers may be employed with advantage for beam delivery in laser Doppler anemometers, especially when high spatial resolution of the measurements is required [134, 135]. Multimode fibers allow the transfer of significantly higher power into the LDA detection volume and need less alignment effort than do the single-mode fibers. High-power laser diodes can be applied in such LDA setups, allowing sensitive velocity measurements of fluid flows. Moreover, the detection volume is smaller than the volume of intersection of the two laser beams because of the low spatial coherence of the multimode light. Thus spatially better resolved measurements of variations in velocity in flows can be achieved. This allows precise frequency measurements and the determination of movements of accelerated particles. The speckle pattern of the multimode beam can be strongly suppressed by choosing high-aperture fibers with high intermodal dispersion and by the use of laser diode arrays with low coherence lengths. The use of multimode-fiber LDA allows achieving higher accuracy in the determination of velocity gradients in laminar and turbulent boundary layers.

Use of the frequency division-multiplexing (FDM) technique in laser-Doppler velocity profile sensors enables the discrimination of signals from two fringe systems [136]. In this case, the sensors utilize only one wavelength and, therefore, the dispersion effect caused by different wavelengths in fiber-optic systems can be avoided. The use of fiber optics separates the optoelectrical part from the measurement head, and thereby improves the robustness of the sensor system. FDM sensors are efficiently used to measure the velocity distribution close to the wall with a high spatial resolution. In addition, the heterodyne technique is frequently used to enable the measurements of small velocities close to zero near the wall [137].

Further in this chapter, we shall discuss two extreme types of applications of laser Doppler velocimetry: long-range velocity measurements with Doppler lidars for atmospheric studies and short-range velocity measurements with laser Doppler microscopes for biomedical studies.

6.2 Long-Range Velocity Measurements and Wind Lidars

Two basic approaches exist in large velocity measurements: coherent and non-coherent determination of the flow-induced Doppler shift frequencies in the radiation scattered by the particles and detected by the lidar. Coherent detection implies that optical mixing is performed of Doppler-shifted scattered and nonshifted reference (or local oscillator) coherent radiation. In this case, the Doppler shift frequency is determined by processing the difference signal in the low-frequency range, and so velocities in a wide range from quite small to very large can be measured. Noncoherent detection implies that the Doppler-shift frequency is determined in the optical frequency range. Certainly, in this case only very large Doppler shifts can be resolved even with modern high-resolution optical spectrometers, so that systems of this type are designed for measuring only large velocities.

The problem of remote measurement of wind velocities is of utmost importance for a variety of basic studies, in particular, in the field of atmospheric physics, and for various applications starting from environmental pollution monitoring and aircraft/airport flight security maintenance to local and global weather forecast and hurricane monitoring. Because in this case the measurements are performed in open atmosphere, the scattering particles are the atmospheric aerosols comprising water droplets, ice crystals, dust, ashes, and other types of particles including those of biological origin (grains, spores, etc.). These particles have different light scattering cross-sections. Thus, depending upon the measurement site and location of the detection volume, the contribution of different particles to the signal significantly varies.

Laser velocimeters for long-range measurements are usually referred to as lidars, the abbreviation "lidar" standing for LIght Detection And Ranging. Depending on the applications, the wind lidars should be operated from a stationary, ground-based laboratory, mobile track, airborne or spaceborne environment. This implies different constraints and limitations in system size, power supply sources, cost, reliability, etc. Doppler lidars that have been developed starting from 1980s are based typically on gas CO_2 (10.6 μm) lasers, diode-pumped solid-state (1.06, 1.55, and 2.0 μm) lasers, and diode-pumped fiber (1.55 μm) lasers.

In systems based on high-power CO_2 lasers, both continuous wave (CW) and pulsed transversely excited atmospheric (TEA) CO_2 sources were used depending on the range of distances to be covered and spatial resolution of velocity measurements to be achieved [138]. In addition to wind velocity magnitude, the systems allowed estimating the value of the velocity structure constant. The method has been experimentally verified in experiments aimed at detecting artificially generated vortexes, e.g., aircraft traveling vortexes, and distinguishing them from natural winds and turbulence based on a comparison of the magnitudes of the velocity structure constant.

Further experiments comparing lidar measurement results with data from conventional Rumbo anemometers and performed from ground-based, sea-based, and atmospheric-mast-based platforms showed that Doppler lidars can measure wind profiles in the atmosphere with excellent accuracy. Invisible atmospheric inhomogeneities such as aircraft wake vortices can be detected and tracked. Further lidars were proposed for spaceborne applications. The backscatter lidar technology experiment (LITE) was successfully tested in space in 1994. A Doppler lidar in space can give both wind and backscatter information. Examples of the proposed spaceborne lidars are discussed in [139, 140].

A prototype of a high-energy, long-pulse, and narrow-bandwidth pulsed CO_2 laser suitable for a spaceborne Doppler wind lidar application is described in [141]. The output energy of 10 J was obtained at greater than 8% efficiency in long, narrow-bandwidth, single-longitudinal, and transverse-mode pulses. A positive branch unstable resonator with a fourth-order super-Gaussian mirror was used as the output coupler. Experiments were carried out to assess the effect of intracavity hard apertures of different diameters that induce diffractive perturbation of the theoretical field and reduce the transverse-mode selectivity of the cavity. An upper limit to the choice of the mirror soft radius has been found, which allows optimization of the trade-off between laser efficiency and beam quality. A value of 0.75–0.8 for the ratio between the $\exp(-1)$ diameter of the beam intensity and the laser clear aperture gave a single-transverse-mode operation without significant loss of efficiency.

In Doppler lidars based on solid state lasers, laser-diode-pumped monolithic ring lasers typically serve as the master oscillator. The oscillators with output power around 2–3 kW are capable of detecting signals from moving clouds at a range of 2.7 km and from atmospheric aerosols at a range of 600 m [142]. A somewhat similar system but aimed at the study of laser-induced hydrodynamic flows and not requiring so much energy is described in [143]. Highly sensitive laser Doppler velocimetry featuring the simultaneous measurement of light-scattering objects moving at different velocities and vibration sensing based on Doppler-shifted light-injection-induced intensity modulation in an externally pumped microchip solid-state laser is demonstrated in [144].

The 1.55-μm CW and pulsed coherent Doppler lidar systems using all fiber optic components have attracted attention for remote wind sensing application because of their eye safety, reliability and easy deployment. The design and performance of a simple, multifunction 1.55-μm CW and frequency-modulated CW coherent lidar system with an output power of 1 W is presented in [145]. The system is based on a semiconductor laser source plus an erbium-doped fiber amplifier, a polarization-independent fiber-optic circulator used as the transmit–receive switch, and digital signal processing. The system is shown to be able to perform wind speed measurements even in clear atmospheric conditions when the visibility exceeds 40 km. The aerosol measurements indicate the potential to use single-particle detection for wind measurements with enhanced sensitivity. The system can perform range and line-of-sight velocity measurements of hard targets at ranges of the order of several kilometers with a range accuracy of a few meters and a velocity accuracy of $0.1 \, \text{m s}^{-1}$.

Problems of signal-to-noise ratio (SNR) and single vs. multiparticle scattering conditions in long-range wind measurements were studied with a CW laser Doppler wind sensor operating at a wavelength of $1.55 \, \mu\text{m}$ by the same authors [146]. At longer ranges ($>100 \, \text{m}$), the signal conforms closely to complex Gaussian statistics, consistent with the incoherent addition of contributions from a large number of scattering aerosols. As the range is reduced, the probe volume rapidly diminishes and the signal statistics is dramatically modified. At the shortest ranges ($<8 \, \text{m}$) the signal becomes dominated by short bursts, each originating from a single particle within the measurement volume. These single-particle events can have a very high SNR because (1) the signal becomes concentrated within a small time window and (2) its bandwidth is much reduced compared with multiparticle detection. Wind signal statistics at different ranges and for a variety of atmospheric backscatter conditions is presented. Results show that single-particle scattering events play a significant role even to ranges of $\sim 50 \, \text{m}$, leading to results inconsistent with complex Gaussian statistics.

In some coherent laser Doppler systems, master oscillator–power amplifier arrangements are used, in which the master oscillator is an external cavity semiconductor laser and the power amplifier is an erbium-doped fiber amplifier with $\sim 1 \, \text{W}$ output at the wavelength of $1.55 \, \mu\text{m}$ [147]. The beams are routed within single mode optical fibers, allowing modular construction of the optical layout with standard components. In a bistatic configuration, separate transmitting and receiving optics are used which ensures sufficient sensitivity for reliable Doppler wind speed detection in moderate scattering conditions at short range (to as much as $\sim 200 \, \text{m}$). The bistatic arrangement leads to a well-defined probe volume formed by the intersection of the transmitted laser beam with the virtual back-propagated local oscillator beam. This can be advantageous for applications in which the precise localization of wind speed is required (e.g., wind tunnel studies) or in which smoke, low clouds, or solid objects can lead to spurious signals. The confinement of the detection volume also leads to a reduction in the signal power.

Use of a laser transmitter based on diode-pumped Ho:Tm:LuLiF, a recently developed laser material that allows more efficient energy extraction, recently enabled building a coherent Doppler lidar at $2 \, \mu\text{m}$ wavelength with a higher output energy ($100 \, \text{mJ}$) than previously available [148]. Single-frequency operation is achieved by a ramp-and-fire injection seeding technique. An advanced photodetector architecture is used, incorporating photodiodes in a dual-balanced configuration. A digital signal processing system allows real-time display of wind and aerosol backscatter data. The high pulse energy and receiver efficiency provide for measurement of wind fields at very long ranges.

6.3 Laser Doppler Microscopes

Laser Doppler microscopes are a class of velocimeters making use of very short focus optic to form a very small detection volume inside the studied object. Usually, these devices are designed for biomedical research, in particular, protoplasmic streaming velocity measurements inside living cells and blood flow velocity measurements in single microvessels. In these applications, the scattering particles are natural constituents of protoplasm and blood, and their sizes, concentration, and optical properties cannot be changed to optimize the measurement conditions. Scattering from the cell and vessel walls and, in some applications, the surrounding medium cannot be totally eliminated. However, it may serve as reference signal, often making the experimental arrangement simpler. Certainly, when experimenting with living objects special precautions should be taken not to affect the specimen and studied flow phenomena with probing light, which might induce heating or specific photoreactions.

Seminal publications showing possibilities of laser Doppler measurements of protoplasmic streaming velocity in two types of cells were made simultaneously [149, 150]. In the experiments with plant cells, the spectrum of light scattered from particles in the streaming protoplasm of a living cell and shifted in frequency by the Doppler effect was measured and interpreted to infer details of the velocity distribution in the protoplasm. The results obtained in the experiments with the fresh-water algae Nitella indicated a characteristic flow pattern to which diffusion makes a negligible contribution. No difference in the velocity of particles of different sizes was found. The streaming velocity varied linearly with temperature with a supraoptimal temperature of $34\,°C$, and the velocity distribution became narrower at high temperatures. The protoplasmic streaming can be inhibited by laser light, and this effect has been used to study the photoresponse of the algae. With beam diameters of about $50\,\mu m$, the inhibition is very local, becoming minimal at a displacement of about $200\,\mu m$ upstream and $400\,\mu m$ downstream. Prolonged exposure produces a bleached area free of chloroplasts, which is 3 orders of magnitude less sensitive to photoinhibition.

Recent studies of the spatial pattern of cytoplasmic streaming velocities performed with the LDV technique in similar live algae cells of Chara were reported in [151, 152]. The LDV method proved to be precise, and yielded reproducible results even when tiny differences in velocities around $20\,\mu m\ s^{-1}$ were measured. A typical Fourier spectrum of the signal containing a well-pronounced Doppler peak is shown in Fig. 6.1. Measurements revealed large spatial and temporal variation in streaming velocity within a cell, independent of the position of the cell with respect to the direction of gravity. In most of the horizontally positioned cells the velocities of acropetal and basipetal streaming, measured at opposite locations in the cell, differed significantly. In the apical parts of the streaming regions of both cell types, acropetal streaming was faster than basipetal streaming. The authors speculate that the endogenous difference in streaming velocities in both rhizoids and protonemata may be

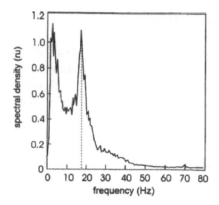

Fig. 6.1. Typical Fourier spectrum of the scattered laser light passing tangentially through the cytoplasm of a Chara cell. Duration of measurements, 1 min. The spectral density is given in relative units (ru). Hz corresponds to $1.22\,\mu m\,s^{-1}$; i.e., the Doppler peak (*dashed vertical line*) corresponds to a streaming velocity of $21.5\,\mu m\,s^{-1}$ [152]

caused by differences in the cytoskeletal organization of the opposing streams and/or loading of inhibitors (such as Ca^{2+}) from the apical/subapical zone into the basipetally streaming endoplasm.

Conventional applications of laser Doppler anemometry and microscopy are related to single scattering limit, which implies that while there may be one or several scattering particles in the detection volume at a time, most photons exhibit only one scattering event during their interaction with the flow. Additional scattering by the neighboring static medium introduces an ambiguity in the direction of the wave vectors of the probe beams and, therefore, a bias in the measured Doppler frequency. As the number of scattering events increases, this bias becomes larger. It is manifested by a broadening of the Doppler peak in the spectrum and by the appearance of additional frequency components, which make the Doppler peak asymmetrical.

Multiple scattering strongly affects the precision in laser Doppler measurements of fluid flow velocities at a high concentration of scattering (tracer) particles in the flow. It makes both the performance of the measurements and interpretation of the data difficult. In particular, this is typical of blood flow measurements in single vessels of diameters higher than $100\,\mu m$ [153,154]. Performance of correct Doppler measurements becomes still more difficult when the flow is imbedded into a strongly scattering medium. This case is typical, e.g., of the measurements of blood flow parameters in the sub-superficial vessels that are imbedded into the tissue, the latter acting as a turbid medium. In conditions of multiple scattering when the mean transport path length of a photon is much shorter than the characteristic size of the studied object, the application of laser Doppler microscopy is not feasible at all. In this case, the approach developed within the framework of correlation theory in diffuse approximation should be used [155,156].

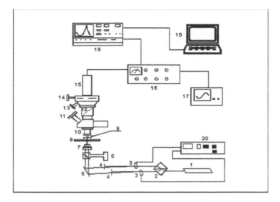

Fig. 6.2. Layout of the experimental setup. 1, He–Ne laser (632.8 nm); 2, beam splitter; 3, Bragg cells; 4, pinholes; 5, tunable mirror; 6, illuminator; 7, focusing lens; 8, XYZ adjustable stage; 9, object under study; 10, collecting lens; 11, oculars; 12, removable mirror; 13, eye piece; 14, tunable pinhole; 15, photomultiplier tube; 16, amplifier; 17, oscilloscope; 18, real-time spectrum analyzer; 19, PC with ADC unit (type L-305); 20, Bragg cell pumping unit [157]

The effect of multiple scattering on the Doppler spectra was studied in test experiments in which a homemade, dual-beam laser Doppler microscope (LDM) operating in forward scattering mode was used to measure the axial flow velocity of a suspension of latex particles in a horizontal glass capillary imbedded into a solution of Intralipid at different concentrations [157]. The layout of the microscope is presented in Fig. 6.2.

In this microscope, acousto-optic frequency shifting is used to split the initial laser beam into two beams of equal intensity and introduce very precise and different frequency shifts. The beams are diffracted in the Bragg cells so that in the first order of diffraction they acquire a tunable relative frequency shift f_{12} from 0 to 10 kHz. The exact value of f_{12} is chosen depending on the maximum velocity to be measured and, consequently, on the maximum Doppler shift to be detected. Typically, f_{12} was set up in the range from 1 to 2 kHz. The intersection of the frequency-shifted and focused probing beams forms the detection volume centered at the axis of the Poiseuille flow in the capillary. Light carries the information about the flow velocity. To be detected, light scattered from the particles traversing the probe volume is focused onto the image plane of the microscope where an adjustable pinhole limits the stray light and cuts out a portion of light from the image of the detection volume. The scattered light that passes through the pinhole is detected by a photomultiplier tube (PMT). The information on the flow velocity is contained in the frequency modulation of the PMT output signal. The power spectrum of this signal normally comprises a low-frequency pedestal and a Doppler peak as shown in Fig. 6.3.

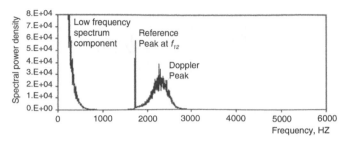

Fig. 6.3. Typical Doppler spectrum obtained from the photomultiplier tube signal of a forward-scattering-mode, dual-beam laser Doppler microscope with acousto-optic frequency shifting of the probing beams when measuring the velocity of a diluted latex particle suspension flow in a thin glass capillary [157]

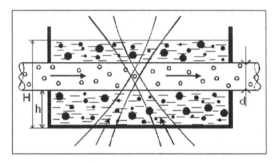

Fig. 6.4. Schematic representation of the measuring cuvette and two probing beams impinging from below the cuvette and forming the detection volume at their intersection [157]

If the measurements are performed in the single-scattering limit, then the Doppler peak is usually centered at the so-called Doppler frequency, proportional to the flow velocity. In this case, the Doppler frequency can be determined from the Doppler spectrum as the frequency that corresponds to the maximum or to the first moment of the Doppler peak, measured relative to the zero frequency. When the relative frequency shift f_{12} is introduced into the probing beams, the Doppler frequency is measured relative to f_{12}, the latter being considered as a reference peak. When the contribution from multiple scattering is high enough, the Doppler spectra become additionally broadened and asymmetrical. In this case, the determination of the flow velocity as proportional to the frequency corresponding to the first moment of the Doppler peak introduces a bias. To evaluate this bias and to explore the limiting values of the scattering parameters upto which the LDM measurements are still feasible, the experiments were performed with a test couvette schematically shown in Fig. 6.4.

A glass capillary with 0.7 mm inner and 1.0 mm outer diameter is imbedded into a Petri dish at height $h = 1.0$ mm over the dish bottom in the plane of the probing beams. The intersecting beams form a probe volume of size $\Delta x \times \Delta y \times \Delta z = 5 \times 5 \times 10\,\mu m^3$ with interference fringes. Further reduction of this volume by a factor of 0.8 is performed by the pinhole placed at the image plane of the microscope. The Petri dish is filled with Intralipid solutions of variable concentrations up to a height $H = 5\,mm$. At $H \leq 3\,mm$, the photons that might have been initially scattered by Intralipid vesicles at depths $z \leq h$ outside the capillary and then by the latex particles inside the capillary do not undergo further scattering by the Intralipid particles. At $H > 3\,mm$, the Doppler-shifted light may undergo additional scattering in the layer of Intralipid solution located above the capillary.

The PMT signal is computer-processed, yielding the current power spectra comprising 4,096 harmonics at a distance of 2.44 Hz in a frequency range 0–5 kHz. The sampling time is 0.41 s. The spectrum processing includes averaging over n current spectra. Figure 6.5 shows the dependence of the spectrum smoothness upon the number of averages. For further spectrum processing,

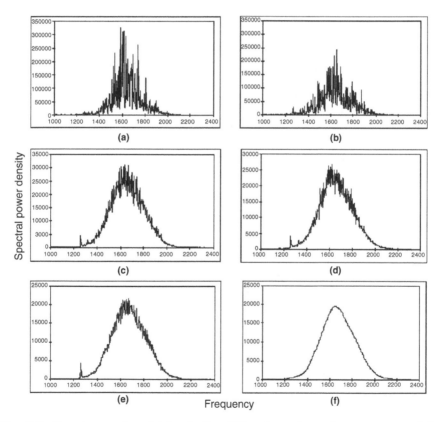

Fig. 6.5. Doppler spectra obtained after different numbers of averages: $n = 1$ (**a, b**), 50 (**c**), 100 (**d**), and 200 (**e**); the spectrum (**f**) was obtained by smoothing the spectrum (**c**) over 21 points [157]

e.g., calculation of the first and the second moments of the Doppler peaks, relative broadening of these peaks, defined as full width at the half-height (FWHH) divided by the frequency of the first moment, and the signal-to-noise ratio, the number of averages is usually chosen depending on the signal properties (typically $n \geq 50$). In addition, the narrow reference peak at the shift frequency f_{12} is subtracted, after which the spectra are smoothed by the method of floating averages over 10–30 points.

6.4 Doppler Optical Coherent Tomographs

In addition to imaging the morphological structure of tissues, functional optical coherence tomography (OCT) techniques allow imaging the vessels with flowing light scattering media, e.g., blood in biological tissue, and measuring the flow parameters. Doppler OCT (DOCT) is an emerging imaging modality that provides subsurface microstructural and microvascular tissue images with near-histological resolution and with a velocity sensitivity less than $1\,\mathrm{mm\ s^{-1}}$ [158–161].

To facilitate understanding of the basic principle, let us describe a simple open air DOCT system, the layout of which is schematically shown in Fig. 6.6 [162].

The low-coherence interferometer of the DOCT system is based on the free-space Michelson interferometer. The light source consists of a superluminescent diode (SLD) with a central wavelength of 822 nm. The full width at the half-maximum (FWHM) of the light source spectrum is 22 nm. These parameters limit the setup's depth resolution to 14.7 µm. The light emitted

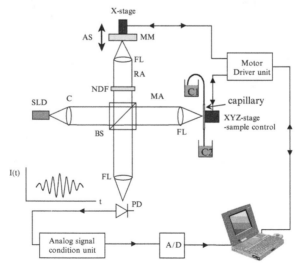

Fig. 6.6. Schematic layout of the open-air DOCT system [162]

by the SLD is first collimated by a collimator (C) and divided into a measurement arm (MA) and a reference arm (RA) using a 50/50 beam splitter. To obtain maximum fringe visibility, the RA also contains a neutral density filter (NDF) to limit the intensity of the reference signal. The reference mirror is placed on a precision translation stage that performs the depth scan with a velocity of $12\,\mathrm{mm\,s^{-1}}$. This velocity gives a Doppler beat frequency of the reference scan of $29.2\,\mathrm{kHz}$. The glass capillary is placed on an XYZ translation stage, which uses stepper motors to move the capillary to the right location. Focusing lenses (FL) with a focal length of $65\,\mathrm{mm}$ and a diameter of $10\,\mathrm{mm}$ are placed adjacent to the RA, MA, and the photodetector (PD). These yield a depth of focus of $87\,\mathrm{\mu m}$ (twice the Rayleigh area), while the spot size, which sets the lateral resolution of the setup, is $6.7\,\mathrm{\mu m}$. The optical signal is converted into an electrical signal by a standard PIN diode. The PD current is converted into a voltage using a transimpedance preamplifier and amplified by means of voltage amplifiers. The bandwidth of the analog signal condition unit is 27–$110\,\mathrm{kHz}$. Further, the signal-to-noise ratio of the amplifier's pass band is $72\,\mathrm{dB}$, limited by the analog to digital (AD) converter. The analog signal is stored on the hard disk of a computer for further signal processing using a 12-bit AD converter at the $300\,\mathrm{kHz}$ sampling frequency.

The flow velocity, v, for each section of the capillary is defined using the equation $v = \frac{(f_\mathrm{D}-f_\mathrm{R})\lambda}{2n_\mathrm{m}\cos(\theta)}$, where λ is the operating wavelength of the SLD, θ is the angle between the SLD beam and the velocity vector of the flow, f_D is the measured Doppler frequency for the section, f_R is the Doppler frequency obtained from the movement of the reference mirror ($29.2\,\mathrm{kHz}$) and n_m is the mean refractive index of the medium.

The outer diameter of the capillary was $1.50 \pm 0.01\,\mathrm{mm}$, while its lumen diameter was $1.01 \pm 0.01\,\mathrm{mm}$. Here, $\alpha = 90°$ and the Intralipid solution is stationary within the capillary.

The standard deviation for the flow velocity along the capillary diameter is 24%. The standard deviation of the flow velocity in the first third of the capillary (0–$330\,\mathrm{\mu m}$) is $0.010\,\mathrm{m\,s^{-1}}$, while the corresponding numbers for the second (330–$660\,\mathrm{\mu m}$) and the last third (660–$1,000\,\mathrm{\mu m}$) are 0.024 and $0.015\,\mathrm{m\,s^{-1}}$, respectively. The increase in deviation is due to the spread of the power at the center of the capillary. This is clearly visible in the Doppler spectrogram shown in Fig. 6.7, which presents a single scan along the axis of the capillary.

The capillary walls are visible as bright dots on the y-axis at $29\,\mathrm{kHz}$. The capillary lumen can be approximately located between 0.27 and $0.40\,\mathrm{s}$ of the scan. The center of the capillary is located at $0.34\,\mathrm{s}$. A comparison between signals recorded near the walls and those at the center reveals that the signal power spreads over a larger frequency range at the center than near the walls. Figure 6.8 illustrates a 3D profile of the flowing 0.3% Intralipid solution.

High-spatial-resolution measurements of light scattering fluid flow velocity profiles with DOCT technique were also reported in [163] in both circular and

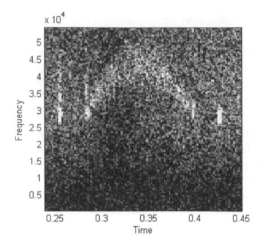

Fig. 6.7. Doppler spectrogram of a single scan of a flowing fluid [162]

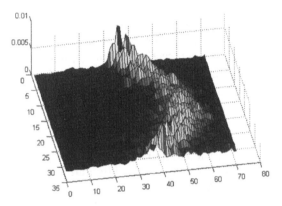

Fig. 6.8. Three-dimensional representation of the velocity profile of 0.3% Intralipid solution flow in a glass capillary [162]

square glass and plastic ducts infused with a moving suspension of microsized particles in water. The ducts were studied both in air and buried in a turbid medium. Although the intensity of backscattered light decreases exponentially when scanning through a turbid medium, the velocity profiles can be clearly resolved because the collected signal is almost entirely due to Doppler-shifted backscattered light from flowing particles within the coherence detection volume. As a result, a substantially higher spatial resolution of the flow velocity can be obtained compared to that of conventional laser Doppler flowmetry.

Potentialities of the DOCT technique for 2D velocity mapping of highly scattering fluids in flows with complex geometry, e.g., a converging flow, are shown in [164]. A complex geometry flow is scanned with a fiber-optic DOCT

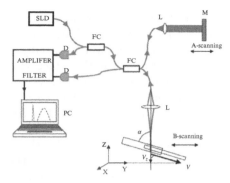

Fig. 6.9. Schematic diagram of the experimental setup: SLD, broadband source; FC, fiber circulators; L, collimating and focusing lenses; M, reference arm scanning mirror; D, dual balanced detectors; and PC, computer [164]

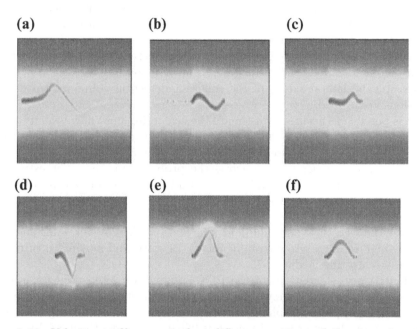

Fig. 6.10. Velocity profiles acquired at different positions of the channel across the center of the tube: (a) before the entry, (b)–(f) after the entry. The horizontal axis represents the depth ranging linearly from 0 (*left*) to 2.2 mm (*right*) for the A-scans through the central line of the channel, while the vertical axis indicates the measured Doppler frequency shifts ranging linearly from −25 kHz (*bottom*) to 25 kHz (*top*) (after [164])

system, as shown in Fig. 6.9, with approximately $10 \times 10 \times 10 \, \mu m^3$ spatial resolution to obtain specific velocity field images.

Concave, blunted, parabolic, and triangular profiles were obtained at different distances after the entry of the flow (Fig. 6.10).

These results show that the DOCT technique may be potentially useful for the study of complex geometry flows in both industrial and biomedical applications, e.g., to study blood circulation, especially in branching vessels. For example, measurements performed at the junction of T- and Y-shaped samples and blood vessels with aneurism show that at constant input volume flow rate, the stationary distribution of flow velocities measured along a cross-sectional plane orthogonal to the inlet arm located at 20 mm off the junction is nonuniform along a junction cross-section plane [165].

The magnitude and broadening of the Doppler shift in DOCT measurements are dependent on a number of factors such as the concentration and flow velocity of the light scattering particles, position and size of the coherent detection volume, and numerical aperture and orientation of the probe beam [163]. Because the particles having higher flow rates remain in the detection volume for shorter durations, an increase in the spectral width of the Doppler peak takes place at higher velocities for equivalent coherent detection volumes. By adjusting the position of the reference mirror, the focus of the probe beam (e.g., 5 μm diameter spot size when using a microlens with NA = 0.22) can be tracked and the coherent detection volume near the focal point can be localized. Such adjustment allows increasing the sensitivity and decreasing the lateral extent of the coherent detection volume when deeper positions are probed.

It is possible to accurately estimate the scattering fluid flow velocity without a priori knowledge of the Doppler angle by combining the value of the Doppler shift in the interference signal and the Doppler spectrum broadening caused by the particles moving across the probe beam [166]. The estimated values of the Doppler angle and average fluid velocity obtained from the experiments agree well with the preset values.

Combining DOCT and OCT techniques in one system, it is possible to visualize the structural and fluid flow information, thereby obtaining simultaneously high-resolution tomographic images of static and moving constituents in highly scattering media [167].

Monte Carlo simulations of the DOCT signal are often used for evaluation of the potential accuracy of DOCT measurements and the inherent noise levels. For example, for a phantom representing the blood flowing in a horizontal 100-μm-diameter vessel placed at 250 μm axial depth in 2% Intralipid solution, numerical calculations demonstrated an accuracy of 3–4% for the depth profiles through the center of the vessel in Doppler frequency values and in position localization of the flow borders, compared with the preset values [168]. Stochastic Doppler frequency noise was experimentally observed as a shadowing in regions underneath the vessel, which was also manifested in simulated Doppler frequency depth profiles. Monte Carlo simulations showed that this Doppler noise has a nearly constant level over about 100 μm underneath the vessel. The noise level is essentially independent of the numerical aperture of the detector and of the angle between the flow velocity and the direction of observation, given this angle is larger than 60°.

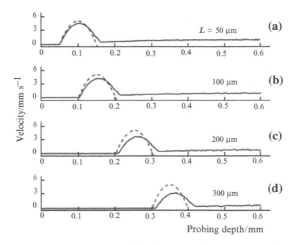

Fig. 6.11. Comparison of Poiseuillie's (real) profile (*dashed curves*) and the velocity profiles reconstructed from the Doppler OCT signal simulated by the Monte Carlo method (*solid curves*) in a blood flow imbedded into a scattering medium at different depths L of flow location

Possible bias in the DOCT signals due to multiple scattering effects was investigated, and apparent velocity profiles for blood flows with a priori specified parabolic velocity profiles imbedded into a light scattering suspension of lipid vesicles at various concentrations were obtained [169, 170]. Monte Carlo simulations showed that the maxima of the reconstructed velocity profiles at high concentrations shift with respect to the symmetry axes of the flow in the direction of the distant boundary and their magnitude decreases, which means that the apparent velocity is lower than that specified in the model, as shown in Fig. 6.11, because of a greater contribution from multiply scattered photons. At the model parameters specified in the simulation, this shift achieves about 18 µm at the flow location depth of 300 µm.

Similar results for a flow phantom consisting of a glass capillary containing whole blood flowing under laminar conditions submerged in a variable depth of the Intralipid solution simulating a blood vessel within the cutaneous microcirculation were obtained experimentally in [171].

Because the DOCT technique has the potential of measuring velocity profiles and their temporal changes with both high temporal and spatial resolutions in blood microvessels with diameters of 100 µm and below, it can be efficiently used to analyze cerebral hemodynamics and its change following, e.g., neural activation. In [172], the cross-sectional profiles of blood flow velocity in the rat pial microvessels and their temporal changes were measured in vivo with an axial resolution of 11 µm and lateral resolution about 14 µm in the cortical tissue. The velocity distributions along the vertical diameter of pial microvessels in a cranial window of the rats were measured at short time intervals by scanning the sampling point repeatedly. The velocity profiles obtained in the pial arterioles were parabolic at any phase, although the centerline velocity pulsated following heart beats with amplitude as large as 50% of the temporal mean velocity. It indicates that the blood flow in the pial microvessels is a quasi-steady laminar flow, which is consistent with the flow expected for the case of a small Reynolds number and a small frequency parameter. The stimulus-induced increase in velocity pulsation was much larger than the increase in the mean velocity, which places a restriction on the mechanism of regulating the regional cerebral blood flow and blood volume.

The implementation of time domain DOCT for blood flow measurement may be limited by low sampling rate [173]. Application of Fourier domain (FD) OCT technique, which has much higher imaging speed [174,175], makes it possible to use DOCT for in vivo blood flow measurements in addition to detection of vessel pulsations. FD DOCT can be used for clinical monitoring of in vivo human skin blood flow by measuring the flow speed not only in the direction of the scanning beam but also in the direction perpendicular to the scanning beam [176]. The reported applications of this method for retinal flow measurements [177] yielded not only linear speed values (mm s^{-1}), but also the volumetric flow parameters (ml min^{-1} or l min^{-1}) without making any assumptions about anatomic features.

Another modification, a moving-scatterer-sensitive optical Doppler tomography (MSS-ODT) technique for in vivo blood flow imaging in real time based on spectral-domain OCT system is proposed in [178]. In MSS-ODT, the influence of stationary scatterers is suppressed by subtracting adjacent complex axial scans before calculating the Doppler frequency shift. MSS-ODT is a useful technique for accurate determination of blood vessel size. The flow profile obtained with MSS-ODT yields a substantially more accurate vessel diameter than that obtained with the conventional phase-resolved method, which underestimates the diameter by about a quarter.

As mentioned above, a specific feature of OCT is its shallow (1–3 mm) penetration depth. This fundamentally limits DOCT imaging to transparent, near-surface, intravascular, or intracavitary anatomical sites. Consequently, interstitial DOCT was developed for minimally invasive in vivo imaging of microvasculature and microstructure at greater depths, providing access to deep-seated solid organs. The feasibility of in vivo real-time visualization and monitoring of the bidirectional blood flow deeply situated in rat leg and abdominal cavity [179] as well as microvascular changes caused by photodynamic therapy (PDT) deep within a tumor in the rat model of prostate cancer using a linear-scanning, needle-based DOCT system was reported in [180]. In particular, a reduction of detected vascular cross-sectional area during treatment and partial recovery post-treatment were observed, microvascular shutdown occurring at different rates and showing correlation with PDT total irradiance and irradiance rates [181]. This capability of the technique may play an important role in elucidating the mechanisms of PDT in tumors, pretreatment planning, feedback control for treatment optimization, determining treatment end points, and post-treatment assessments.

Endoscopic in vivo applications of DOCT can be facilitated by integration with a small ($1.4 \times 1.0\,\text{mm}^2$) elliptical microelectromechanical system (MEMS) membrane mirror, electrostatically actuated to dynamically adjust the optical beam focus and track the axial scanning of the coherence gate in the DOCT system [182]. The MEMS mirror is designed to maintain a constant numerical aperture of ~ 0.13 and a spot size of $\sim 6.7\,\mu\text{m}$ over an imaging depth of 1 mm in water at a scanning frequency of 8 kHz, which improves imaging performance in resolving microspheres in gel samples and Doppler shift estimation precision in a flow phantom.

Clinical feasibility of an endoscopic DOCT (EDOCT) system in the human gastrointestinal (GI) tract was assessed in [183]. During routine endoscopy, patients were imaged with a prototype EDOCT system, which provided color-Doppler and velocity-variance images of mucosal and submucosal blood flow at one frame per second, simultaneously with high-spatial-resolution ($10–25\,\mu\text{m}$) images of tissue microstructure. Images were acquired from both normal GI tract and pathologic tissues. Differences in vessel diameter, distribution, density, and blood-flow velocity were observed among the GI tissue pathologies imaged. EDOCT could detect different microcirculation patterns exhibited by normal and diseased tissues, which may be useful for diagnostic imaging and treatment monitoring.

Color Doppler OCT (CDOCT) is an innovation that enhances the spatially resolved flow-velocity mapping simultaneously with microstructural imaging. However, stochastic alterations of the Doppler spectrum by fluctuating distributions of the scattering particles in the flow field give rise to unavoidable inaccuracies in velocity estimation as well as to a fundamental trade-off between image acquisition rate and velocity precision. Algorithms that permit high-fidelity depth-resolved measurements of velocities in turbid media have been reported in [184].

In some specific, in particular biological, applications, imaging of dynamic samples can be also performed using spectral domain phase microscopy (SDPM), which is basically a functional extension of spectral domain optical coherence tomography [185]. SDPM achieves exquisite levels of phase stability by employing common-path interferometry. Like in OCT, axial resolution in SDPM is determined by the source coherence length, while lateral resolution is limited by diffraction in the microscope optics. Nonetheless, the quantitative phase information SDPM generates is sensitive to sub-Angstrom displacements of the scattering structures. Integrative quantitative phase imaging techniques, such as Fourier phase microscopy, Hilbert phase microscopy, and digital holographic microscopy, have achieved sub-micron motion detection in live cells. In contrast with these techniques, SDPM can achieve full depth discrimination, allowing the resolution of the motion of independent, sub-cellular structures at various cross-sectional planes within the sample. This enables the researchers to apply this technique to monitor the thermal contraction of a glass sample with nanometer per second velocity sensitivity, and to measure the cytoplasmic streaming in live cells, e.g., in an Amoeba proteus pseudopod. The authors observed a reversal of the cytoplasmic flow induced by extracellular $CaCl_2$ and a parabolic velocity distribution therein.

In [186], the authors extend the use of SDPM to produce 3D reconstructions of the internal and surface motions of beating cardiomyocytes. Phase information is used to quantify the motion of cellular structures in the axial dimension. The gated acquisition process involves synchronization of the SDPM detection system with an applied electrical field used to stimulate beating in isolated cardiomyocytes. For a given pacing protocol, repeat motion measurements in two dimensions during the cellular contraction are obtained, building a volume image by repeating the process at multiple discrete slices through the cell. This experiment serves as a proof of principle for volumetric imaging of beating cardiomyocytes.

6.5 Laser Doppler Flowmeters and Perfusion Imagers

An excellent introduction to the history, theoretical principles, and applications of laser Doppler flowmetry (LDF) can be found in [187]. Particular applications of LDF to specific tissues include intravascular catheter velocimeters as well as cutaneous, skeletal muscle, gastrointestinal and respiratory tract, central and peripheral nervous system, renal, bone, cochlear, and retinal applications. However, there are many unique features and potential artifacts inherent to this technique that should be taken into account, especially by clinicians and physiologists, when interpreting the measurement data acquired by commercially available laser Doppler flowmeters.

Temporal and spatial heterogeneity of blood flow are real physiological phenomena contributing to the variability of flow measurements in small (around $1 \, mm^3$) tissue volumes. They are manifested by the variability of optical properties of flowing blood and, consequently, by the output signal of any measuring device [188, 189]. Even an ideal technique would demonstrate variations in the measured flow. Superimposed on these physiological variations are the technical artifactual variations of LDF. This is why it has been and is still being paid to experimenting with in vitro perfusion flow models and numerical simulations of the behavior of red blood cell suspensions and undiluted blood samples in flow and stasis [190]. For various shear rates and blood layer thicknesses, the angular light distributions and the collimated transmission were measured for different laser wavelengths. In particular, it was shown that at lower shear rates the total attenuation of red He–Ne laser light followed an irregular pattern. From the angular intensity distributions, the anisotropy for single scattering was deduced by inverse Monte Carlo simulations. A continuous increase of the scattering anisotropy factor, g, with the shear rate was observed, with g in the range 0.95–0.975.

In LDF devices, the measurements are usually performed with measuring heads comprising the emitting (source) and receiving (detector) units, in particular, fiber-optic elements placed at various relative distances (source–detector separations). The receiving unit detects the light scattered by the moving particles, red blood cells, and stationary particles constituting the tissue cells surrounding the blood flow. Coherent optical mixing of these two fluxes on a quadratic detector enables obtaining the spectrum of beat frequencies (Doppler spectrum), which is then processed to estimate the blood flow rate.

While in most conventional laser Doppler instruments small source–detector separations are used, measuring blood perfusion in the deeper regions of tissue requires the use of large separations. The influence of the depth of the perfusion on the Doppler spectrum depends on the path length that the detected light travels at different depths. Hence, for the large source–detector separations this influence is different from that for the conventional LDF instruments, which should be taken into account [191]. Spectral analysis is traditionally used to process the LDF spectra. As one of the new trends in signal processing, application of the wavelet transform should be marked. It is an especially valuable tool in regard to the periodic oscillations of the cutaneous LDF signal representing the influence of heart beat, respiration, intrinsic myogenic activity, and the neurogenic factors on the cutaneous blood flow. These oscillations reflect the state of the microcirculatory system and reflect different pathological processes, stress, and physical-exercise-induced reactions, etc. The wavelet transform gives good time resolution for high-frequency components and good frequency resolution for low-frequency components. LDF measurements combined with wavelet analysis of oscillations in the peripheral blood circulation has a great potential for studies of both the micro- and macroscopic mechanisms of blood flow regulation in vivo [192, 193].

Systems currently being developed for full-field laser Doppler blood flow imaging allow 2D flow maps or monitoring flux signals to be obtained from a plurality of measured points simultaneously by using a 2D array of photodetectors [194, 195]. The detection part of such systems is based on an intelligent complementary metal oxide semiconductor (CMOS) camera [196] with a built-in digital signal processor and memory to detect Doppler signals in a plurality of points over the area illuminated by a divergent laser beam of uniform intensity profile. The imaging time of these systems is several times shorter than for the conventional scanning laser Doppler imager. For example, high-resolution flow images (256×256 pixels) can be delivered every 2–10 s, depending on the number of points in the acquired time-domain signal (32–512 points). The integrating property of the detector improves the signal-to-noise ratio of the measurements, which results in high-quality flow images. The very nice performance of the imager in vivo shows the good prospects for future implementations of the imager for clinical and physiological applications.

CW laser sources are commonly used in laser Doppler flowmeters. When applied to biotissues, the spatial and spectral (in relation to Doppler frequency spectrum) resolution of the CW LDF is low, as a consequence of intense light scattering. Much higher spatial and spectral resolution could be attained by combining a conventional laser Doppler with time-resolved (TR) measurement techniques. Time-resolved laser spectroscopy (TRS) of stationary biological tissues is a rapidly developing method for biomedical tomography and diagnostics. With the time gating technique, it is possible to filter away the photons traveling over a relatively long time along strongly randomized trajectories and to detect only the early arriving photons whose paths just slightly deviate from an average trajectory, connecting the point of incidence of laser light with the point of detection. In [197], the feasibility of measuring heterodyne Doppler shift spectra from a moving, highly scattering medium for three pulsing laser sources: a nanosecond laser diode, a 35 ps duration of pulses Nd:YLF laser, and a 3–8 ps pulse duration dye laser, was demonstrated. The obtained results open the prospects for time-gated LDF for blood perfusion tomography.

In clinical use, because laser Doppler imaging is capable of accurately predicting the traumatic, e.g., wound, outcome with a large weight of evidence, this technique has been approved for burn-depth assessment by regulatory bodies including the Food and Drug Administration (FDA) [198]. In this particular clinical application, the use of telemetry and simple burn photographs may be the best option for the initial emergency assessment. However, for the actual treatment decisions, laser Doppler techniques have been shown to be superior to such perfusion measurement techniques as thermography, vital dyes, video angiography, video microscopy, etc.

Assessment of cerebral blood flow (CBF) changes with a wide-field laser Doppler imager based on a CCD camera detection scheme, in vivo, in mice is reported in [199]. The used setup enables the acquisition of data in minimally invasive conditions. In contrast with conventional laser Doppler velocimeters and imagers, the Doppler signature of moving scatterers is measured in the frequency domain, by detuning a heterodyne optical detection. The quadratic mean of the measured frequency shift is used as an indicator of CBF. A significant variability of this indicator in an experiment designed to induce blood flow changes was observed.

With existing optical imaging techniques, 3D mapping of microvascular perfusion within tissue beds is severely limited by strong scattering and absorption of light by tissue. To overcome this limitation, an effective digital frequency modulation approach to achieve directional blood flow imaging within microcirculations in tissue beds at sub-millimeter tissue depths in vivo for optical microangiography (OMAG) has been developed [200–204]. The method only requires the system to capture one 3D data set within which the interferograms are modulated by a constant frequency modulation that gives one-directional flow information. The result is that the imaging speed is doubled and the computational load is halved. The method is efficiently used for imaging of cerebral vascular blood perfusion in a live mouse with the cranium left intact, or the blood circulations in posterior chamber of the human eye. Use of phase changes in sequential OCT A-scan signals allows minimizing of

the motion artifacts in the flow image caused by the inevitable subject movement. OMAG is capable of providing volumetric vasculatural images in retina
and choroids, down to capillary level imaging resolution, within ~ 10 s.

6.6 Particle Image Velocimeters (Including Capillaroscopes and Angiographers)

Particle image velocimetry (PIV) and especially the more modern digital particle image velocimetry (DPIV) are quantitative high-resolution dynamic techniques that are able to measure instantaneously the 2D and 3D velocity fields
and capture detailed chronological changes in these fields. Many applications
of PIV were featured in the proceedings of the serial International Flow Visualization Symposia, first held in Tokyo in 1977 and then successively in Bochum,
Ann Arbor, Paris, Prague, Yokohama, Seattle, Sorrento, Edinburgh, Kyoto,
Notre Dame, and Göttingen [205–207].

Comprehensive descriptions on the PIV technique including the fundamentals and relevant theoretical background information that directly supports
the practical aspects associated with the performance and understanding
of the experiments employing the PIV technique can be found, e.g., in
[208–213].

The main idea of the method is to measure the velocity by imaging and
tracking the motion of tracer particles suspended in the fluid. The acquired
images of the flow are evaluated by dividing the images into so called "interrogation areas" and cross-correlating the interrogation areas in consecutive images. For each interrogation area, the displacement vector is calculated from
the location of the correlation peak. The velocity field is simply calculated
from the displacement field by dividing the displacement vectors by the time
separation between the images.

The basic requirements for a PIV system are an optically transparent
test section, an illuminating light source (laser), tracer particles, a recording medium (film, CCD, or holographic plate), and a computer for image
processing. Tracer particles for PIV must satisfy two requirements [214]: (1)
they should be able to follow the flow streamlines without excessive slip, and
(2) they should be efficient scatterers of the illuminating laser light. While
the first requirement is fairly obvious, the second requirement significantly
impacts the illuminating lasers and recording hardware. For example, if a
given particle scatters weakly, then one would have to employ more powerful
lasers or more sensitive cameras, both of which can drive up costs, as well as
the associated safety issues. Although the search for ideal particles may seem
somewhat trivial, it can potentially provide immense benefits. In biomedical
applications of PIV, typically the flows of blood and blood mimicking suspensions are studied. In this case, usually the red blood cells play the role of
tracing particles.

When working with low-speed liquid flows, it is beneficial to use particle densities close to the liquid density. For example, polystyrene and other plastic particles that have densities within a few percent of water are good candidates for water flows. For gaseous flows, typically oil droplets are used (silicone oil or organic oils such as olive oil). Owing to the very large density difference between oil and the gaseous medium, it becomes essential to use very small droplets, typically <1 mm, to minimize the settling velocities.

PIV measurements contain errors arising from several sources [215]: (1) random error due to noise in the recorded images; (2) bias error arising from the process of computing the signal peak location to subpixel accuracy; (3) gradient error resulting from rotation and deformation of the flow within an interrogation spot leading to loss of correlation; (4) tracking error resulting from the inability of a particle to follow the flow without slip; and (5) acceleration error caused by approximating the local Eulerian velocity from the Lagrangian motion of tracer particles. Certain errors can be minimized by careful selection of experimental conditions (for example, tracking error). However, other sources are inherent to the nature of the correlation in PIV and cannot be eliminated. For example, even if the recorded images are free from noise, the location of the correlation peak can be influenced by random correlations between particle images not belonging to the same pair. In addition, bias errors result from a phenomenon called "pixel locking", in which the signal peak location is biased towards the nearest pixel while using a curve-fit to locate the discretized signal with subpixel accuracy. Similarly, gradient errors will occur in turbulent flow. Finally, acceleration error cannot be eliminated because of the very principle of PIV, which uses the Lagrangian motion of particles to approximate the instantaneous Eulerian flow velocity. In order to evaluate and, possibly, reduce the measurement error, a detailed analysis should made accounting for the specific features of the experiment, including the studied system, used equipment, and data processing algorithms.

Starting with the seminal work by Maynart [216] who demonstrated the strength of this method by mapping the particle distribution in a Rayleigh–Bernard water-flow visualization exercise, a variety of alternative PIV techniques have been developed, in particular, applying laser light sheet illumination of particle-seeded flows. Successive video images, recorded perpendicular to a light sheet parallel to the main stream, are digitized and processed to map the flow velocity in 2D planes. In particle tracking velocimetry (PTV), displacements of single particles in two subsequent images are determined semiautomatically, resulting in flow diagrams consisting of nonuniformly distributed velocity vectors. Application of grid-cell averaging results in flow field diagrams with uniform vector distribution. In subimage correlation PIV (SCPIV), repetitive convolution filtering of small subareas of two subsequent images results in automatic determination of cross-correlation peaks, yielding flow field diagrams with regularly spaced velocity vectors. In both PTV and SCPIV, missing values, caused by incomplete particle displacement information in some areas of the images or due to rejection of some erroneous vectors by the vector validation procedure, are interpolated using, e.g., a 2D spline interpolation technique. The resultant vector flow fields are used to study the spatial distribution of velocity, spatial acceleration, vorticity, strain, and shear. These flow fields could also be used to test for flow in the third dimension by studying the divergence, and to detect the presence and location of vortices. The results offer detailed quantitative descriptions of the flow morphology and can be used to assess dissipated energy. The versatile character of the technique makes it applicable to a wide range of fluid measurements.

One of the modern trends in PIV is to raise the temporal resolution of measurements. It is especially important in aerodynamic and aerospace application, including transonic flows [217]. Considerable progress has been made during the last 20 years in this field. Although many problems are still to be solved in turbulent flows, mean flow characteristics can be quite accurately defined by the combined use of computer codes and wind tunnel experiments. To improve the efficiency of their aircrafts and engines, the manufacturers are looking to the 3D, unsteady, aspects of the flow in some specific subdomain or in particular flight or working configurations. Complex viscous flows, separated flows, coherent structures in turbulence, and transition phenomena become of primary importance in such situations. In these fields, computer simulations are still limited, and much understanding has to be gained from experiments. Further raising the temporal resolution of digital PIV is one of the most efficient approaches.

In the case of transonic flows, the implementation of digital PIV for a range of free-stream flow speeds from 0.1 to 0.8 Mach number in wind tunnel typically yields a measurement error of about 1%. As an option, in such measurements, a double-pulsed, frequency-doubled Nd:YAG laser with 100 mJ per pulse and 25 Hz repetition rate can be used to form the light sheet. The pulse separation can be varied from 20 ns to 1 ms. The green output wavelength of 512 nm also makes it more optically efficient in scattering light from the styrene seeding particles. The laser firing signal should be synchronized with the vertical framing pulse of the CCD camera. In this manner, both camera and light sheet can be tracked to different measurement stations in the flow field.

Another trend in PIV is to raise the spatial resolution of measurements. For this purpose different variants of micro-PIV techniques are being developed. One of the approaches to directly measure particle velocity and concentration in a sheared suspension flow with submicrometer spatial resolution is to combine the PIV technique with evanescent-wave illumination, as shown in [218]. A 0.1% volume fraction aqueous suspension of 300–500 nm diameter fluorescent polystyrene and silica particles was sheared between parallel stationary glass and rotating textured silicon disks, a flow geometry that models the chemical-mechanical polishing (CMP) process used to plane silicon wafers. The technique was used to visualize only the particles within 300 nm of the glass surface, or only the particles that interact with this surface. PIV was used to determine the in-plane velocity components in this near-wall region. Although limited to the near-wall region, PIV using evanescent-wave illumination has two advantages over other micro-PIV techniques: a finer spatial resolution (50–300 nm vs. a few micrometers), and no extra light since only the region of interest is illuminated.

In another application of the micro-PIV technique, the high-resolution, near-wall fluorescent microparticle image velocimetry was used in mouse cremaster muscle venules in vivo to measure velocity profiles in the red-cell-depleted plasma layer near the endothelial lining [219]. The micro-PIV data of the instantaneous translational speeds and radial positions of fluorescently labeled microspheres (0.47 μm) in an optical section through the midsagittal plane of each vessel were used to determine the particle translational speeds. Regression of a linear velocity distribution based on near-wall fluid-particle speeds consistently revealed a negative intercept when extrapolated to the vessel wall. On the basis of a detailed 3D analysis of the local fluid dynamics, a mean effective thickness of approximately 0.33 μm for an impermeable endothelial surface layer was estimated. The extent of plasma flow retardation through the layer required to be consistent with the micro-PIV data results in near-complete attenuation of fluid shear stress on the endothelial cell surface. These findings confirm the presence of a hydrodynamically effective endothelial surface layer, and emphasize the need to revise the existing concepts of leukocyte adhesion, stress transmission to vascular endothelium, permeability, and mechanotransduction mechanisms.

The hemodynamic characteristics of blood flow are important in the diagnosis of circulatory diseases, since such diseases are related to wall shear stress of cardiovascular vessels. This implies a necessity for various experimental studies with model systems. Chicken embryos are frequently used as such systems. In [220], the micro-PIV technique was employed to assess blood flow in extra-embryonic venous and arterial blood vessels of chicken embryos, using red blood cells (RBCs) as tracers and obtaining flow images of RBCs using a high-speed CMOS camera. The mean velocity field showed non-Newtonian flow characteristics. The blood flow in two venous vessels merged smoothly into the Y-shaped downstream vein without any flow separation or secondary flow. The vorticity was high in the inner regions, where the radius of curvature varied greatly. A periodic variation of temporally resolved velocity signals, due to beating of the heart, was observed in arterial blood vessels. The pulsating frequency was obtained by fast Fourier transform analysis using the measured velocity data. The measurement technique was shown useful in analyzing the hemodynamic characteristics of in vivo blood flow in chicken embryos.

High-resolution PIV technique can be applied to in vivo imaging the arterioles in the rat mesentery using an intravital microscope and a high-speed digital video system [221, 222]. The technique is capable of taking into account the mesenteric motions, thereby enabling the researchers to obtain the velocity distributions with spatial resolutions of $0.8 \times 0.8 \, \mu m^2$ even near the wall in the center plane of the arterioles and with 1 ms time interval. The arteriole velocity profiles appear to be blunt in the central region of the vessel cross-section and sharp in the near-wall region. In certain cases, in single and straight arterioles as well as in bifurcations the velocity profiles exhibited a pit at the central region. These flow features are typical of non-Newtonian fluids.

One of modern applications of digital PIV systems is to examine the flow structure in biomechanical devices with complex geometry, in particular, in mechanical heart valves (MHVs). Hundreds of thousands of MHVs are implanted each year in the US and European countries. Flow through the MHV hinge can cause thromboemboli formation. Examination of various orifice geometries representing the MHV hinge region and how these geometries may contribute to platelet activation and thrombin generation can be efficiently performed with the help of DPIV [223]. The results of conducted experiments indicate that small changes in geometry, although do not affect the bulk flow, change the coagulation propensity as blood flows through the orifices. A more abrupt geometry allows more stagnation to occur resulting in more thrombin generation. DPIV showed differences in the jets with respect to entrainment of stagnant fluid. These results help to pinpoint the important parameters that lead to flow stasis and subsequent thrombus formation.

In another experimental study [224], the evolution of the flow field in a detailed time domain of a commercial bileaflet MHV in a mock-loop mimicking unsteady conditions was studied. The investigated flow field corresponded to the region immediately downstream of the valve plane. The combination of a Nd:YLF high-repetition-rate double-cavity laser with a high-frame-rate CMOS camera allowed a detailed, highly temporally resolved acquisition (up to 10,000 fps depending on the resolution) of the flow downstream of the MHV. Features that were observed include the nonhomogeneity and unsteadiness of the phenomenon and the presence of large-scale vortices within the field, especially in the wake of the valve leaflets. Furthermore, the temporally cycle-resolved analysis allowed the different behaviors exhibited by the bileaflet valve at closure to be captured in different acquired cardiac cycles. By accurately capturing hemodynamically relevant time scales of motion, time-resolved PIV characterization can realistically be expected to help designers in improving the MHV performance and in furnishing comprehensive validation with experimental data on fluid dynamic numeric modeling.

To enhance the noninvasive measurements of velocity profiles, multiple velocity vectors, and shear stress, it is possible to combine PIV with ultrasound contrast imaging [225]. The combined method (echo-PIV) takes advantage of the strong backscatter characteristics of small gas-filled microbubbles (contrast) seeded into the flow. The test measurements conducted in vitro show that the velocity profiles measured by echo-PIV agree well with those measured by optical PIV profiles both in the steady flows and in pulsatile flows. Echo-PIV follows the general profile of pulsatile shear stress across the artery but underestimates the wall shear at certain time points. However, the error in shear from echo-PIV is an order of magnitude lower than that from current shear measurement methods.

Modern trends in applications of the micro-PIV technique, in particular, to study blood microcirculation, are associated with the progress in microfabrication technologies [226]. Microfluidic devices are expected to provide powerful tools not only to better understand the biophysical behavior of blood flow in microvessels, but also for disease diagnosis. Such microfluidic devices for biomedical applications must be compatible with state-of-the-art flow measuring techniques, such as confocal PIV. The confocal system should have the ability not only to quantify the flow patterns inside microchannels with high spatial and temporal resolution, but also to provide the velocity measurements for several optically sectioned images along the depth of the microchannel. For example, when integrated with a square or rectangular polydimethysiloxane microchannel, a confocal micro-PIV system allows imaging the trace particles seeded in the flow and acquiring the velocity profiles by successive measurements at different depth positions to obtain 3D information on the behavior of fluid flows [227, 228]. Like in other examples, the velocity profiles obtained with such system for red blood cell suspensions, are markedly blunt in the central region; however, the exact profile shape depends on the hematocrit.

In [229], the performance of a high-speed confocal micro-PIV system was demonstrated in an application to the internal flow measurement of a droplet passing through a microchannel. The system enabled the researchers to obtain a sequence of sharp and high-contrast cross-sectional particle images at 2,000 frames per second and to measure the velocity distributions of microflows with a confocal depth of $1.88\,\mu$m. Three-dimensional distributions of three-component velocities of a droplet traveling in a $100\,\mu$m (width) $\times\,58\,\mu$m (depth) channel were measured. A volumetric velocity distribution inside a droplet was obtained, and the 3D flow structure inside the droplet was investigated. In particular, the measurement results suggest that a 3D complex circulating flow is formed inside the droplet.

Another modern trend is to integrate several laser-based techniques, in particular, micro-PIV and optical tweezers (OT), thereby allowing manipulation and characterization of the mechanical environment in and around microscale objects in a fluidic environment. The resulting instrument, the micro-PIVOT, enables a new realm of microscale studies, still maintaining the individual capabilities of each optical technique. This was demonstrated in [230], with individual measurements of optical trap stiffness ($\sim 70\,\mathrm{pN}\,\mu\mathrm{m}^{-1}$ for a 20-μm polystyrene sphere and a linear relationship between trap stiffness and laser power) and fluid velocities within 436 nm of a microchannel wall. The integrated device was validated by comparing computational flow predictions to the measured velocity profile around a trapped particle in either a uniform flow or an imposed, gravity-driven microchannel flow ($R^2 = 0.988$, RMS error $= 13.04\,\mu\mathrm{m}\,\mathrm{s}^{-1}$). Interaction between both techniques was shown to be negligible for 15–35-μm diameter trapped particles subjected to fluid velocities from 50 to $500\,\mu\mathrm{m}\,\mathrm{s}^{-1}$ even at the highest laser power (1.45 W). The integrated techniques will provide a unique perspective toward understanding microscale phenomena including single-cell biomechanics, non-Newtonian fluid mechanics, and single particle or particle–particle hydrodynamics.

An efficient integration of different laser techniques for the registration of 3D flow velocities resulted in the development of holographic particle image velocimetry (HPIV) [231–233]. The major step forward made by this technique is to eliminate most of the depth-of-focus restrictions of classical PIV by a holographic recording of tracer particles. Thus, even nonstationary flows can be registered in a single record. Significant advancements toward practical systems were made possible by the recent technological developments of high-energy pulsed lasers and electronic image acquisition, as well as the increasing performance of digital image processing. Modern holographic velocimetry approaches can be grouped into two sections according to the type of hologram recording: using either a physical carrier material or an electronic image sensor. Somewhat anachronistically, the silver-halide emulsion of photographic film is still used especially when high resolving power is needed. It offers still unequalled resolution of up to 5,000 line pairs/mm at reasonable sensitivities to record even the low-power light scattered by tiny tracer particles; yet it requires laborious wet chemical processing.

The huge amount of data that can be stored on a photographic film and the immense effort needed to analyze the reconstructed holographic images are given in [234]. An efficient analysis of hundreds of holograms is achieved with an automated scanning system incorporating an optical bandpass filter to suppress the speckle noise and to detect small tracer particles even in the neighbourhood of much larger particles. Data flow is economized by a new and very efficient compression method.

While the speckle noise is much reduced in the widely used off-axis recording, it still has a considerable impact when large and densely seeded volumes are under investigation. Light-in-flight holography (LiFH) is a way to suppress the noise by coherence reduction. Application of this approach to a large wind tunnel flow is discussed in [235]. While a complete deep-volume field is recorded, the effective depth during the read-out process is considerably reduced from too many out-of-focus particle images. Analysis of the digitized real-image particle field is done by direct 3D correlation of grey values from the depth scans.

Reconstruction from a double exposure hologram permits direct correlation of the complex amplitude of the particle fields locally by the object conjugate reconstruction (OCR) method [236]. This technique is very well suited for measuring within thick glass cylinders (e.g., in engine research) because it dispenses with the need to correct for distortions by use of holographic optical elements. Instead, it uses a ray tracing analysis for correct mapping.

In order to reduce the time delay and efforts in processing the standard photographic films, a genetically modified version of the photochromic protein bacteriorhodopsin (BR) has been recently introduced to HPIV measurements. As shown in [237], the use of BR offers excellent resolution and sufficient optical sensitivity and allows a variety of configurations, in particular, utilizing its ability to alter the state of polarization in the reconstruction. Currently, BR is considered as the most promising candidate to replace the silver halide photographic film, and also for intermediate storage of holograms to be scanned and processed digitally.

To enhance data processing, special purpose computer systems are designed for digital HPIV as described, e.g., in [238]. For calculating the intensity of an object from a hologram by fast Fourier transform, special chips are used that can make 100 reconstructed images from a 256×256 grid hologram in $266 \, \text{ms}$. Such computer systems dramatically improve the efficiency of analysis in digital HPIV.

A specific biological application of the micro-PIV technique was reported recently in [239]. The authors measured with a special PIV program the time-dependent velocity fields of protoplasmic streaming over the whole area of a nondifferentiated cell (plasmodium) of the slime mold Physarum. Combining these data with measurements of the simultaneous movements of the plasmodia has revealed a simple physical mechanism of locomotion. The shuttle streaming of the protoplasm was not truly symmetric because of the peristalsis-like movements of the plasmodium. This asymmetry meant that the transport capacity of the stream was not equal in both directions, and a net forward displacement of the center of gravity resulted. This mechanism may be general for all cells with the amoeboid type of locomotion.

In clinical diagnostic applications, simple, fast, and high-quality imaging of single capillaries and capillary networks is very important. Particular sites on human body where such imaging is relatively easy to perform are the nailfolds of fingers and toes. Nailfold capillary microscopy shows an impressive cost/effectiveness ratio: it is simple, noninvasive, and inexpensive. Video recordings of RBCs flowing through capillary networks contain a considerable amount of information pertaining to oxygen transport through the microcirculation. Image analysis of these video recordings has been widely used to determine the RBC dynamics (velocity, lineal density, and supply rate) and oxygenation. However, not all capillaries in a given field of view are suitable for image analysis. Typically, capillary segments that are relatively straight and in sharp focus and exhibit flow of individual RBCs that are well separated by plasma gaps are good candidates for analysis. Various image processing tools have been and are still being developed to aid in the selection of such capillaries for analysis and to obtain quick overviews of RBC flow through the microcirculation (see, e.g., [240–244]). In particular, it is possible to highlight all capillaries containing RBCs in a given field of view and to identify the capillaries that exhibit high lineal density or stopped flow: or to identify capillary segments that are in good focus and are perfused by RBCs and plasma and indicate the frequency of passage of RBCs separated by plasma gaps. So, the flow visualization techniques are valuable tools to aid in the study of image focus, network geometry, RBC flow paths, and dynamics, which can then be used in identifying capillaries for subsequent detailed analysis to provide quantitative information about RBC flow. In addition to dynamic information, the peripheral microvascular damage can be quantitatively analyzed, which is typical of many diseases and which, in particular, is characterized by increasing structural alterations of the capillaries (giant capillaries and microhemorrhages) with progressive decrease of their density.

To extend the range of measurable velocities up to about $20\,\mathrm{mm\ s^{-1}}$, electronic shuttering of an intensified CCD camera can be used to produce multiple images of a single RBC in the same video frame [245]. Velocity is computed as the product of the distance between centroids of two consecutive image positions of RBC and the shuttering frequency of the camera intensifier. In the experiments performed using this approach in microvessels of the hamster retractor muscle with diameters ranging from 8 to 50 μm, mean RBC velocity profiles were found to be blunter than would be expected for Poiseuille flow. Single RBCs tracked along an unbranched arteriole exhibited significant temporal variations in velocity.

Imaging individual RBCs with micrometer spatial resolution and measuring RBC flow velocities with millisecond temporal resolution can be performed by using two-photon microscopy [246]. For example, using this technique for in vivo imaging the individual RBCs in glomerular capillaries in the rat dorsal olfactory bulb, the authors found that odor stimulation evokes capillary vascular responses that are odorant- and glomerulus specific. These responses consist of increases as well as decreases in RBC flow, both resulting from independent changes in RBC velocity or lineal density. It was demonstrated that in olfactory bulb superficial layers the capillary vascular responses precisely outline the regions of synaptic activation.

To elucidate the function of the microcirculation system, it is very important to know the distribution of blood flow velocities inside the microvessels. It can be done automatically by image correlation [247]. The "window" in the image correlation method is equivalent to the sensors in various other measurement methods. By reducing the size of a circular window to the size of erythrocytes it is possible to measure in vivo blood flow velocities with high accuracy. In particular, recording the images with a high-speed video camera system at high temporal resolution, it is possible, for instance, to identify the sites in the vessels where the erythrocytes flow faster and the vessel wall is exposed locally to higher shear stress in the hypertensive condition, which can cause further development of a pathology.

Another example is studying of the phenomenon blood flow structuring, i.e., coordinated self-organization of RBCs in their flow along the microvessels, which actually defines the blood rheological properties in their lumina [248]. These studies are very important because many local hemorheological disturbances in the microvessels are related to intensified RBC aggregation and to the subsequent local accumulation in the microvascular lumina, thereby entailing disorders of the blood flow structuring. White blood cells and platelets are not directly involved in the normal blood flow structuring in microvessels, but they can largely affect both the blood flow normal structuring and the flow velocity under various pathological conditions.

6.7 Conclusions

There is a great demand for velocity measurement techniques and the number of applications is increasing dramatically. Nondisturbing flow velocity monitoring and control has many applications in natural, industrial, and physiological environments. Basic research is also in need of reliable velocimeters for various applications in physics, mechanics, chemistry, biology, fundamental medicine, and what not. Clinical diagnostic measurements need new faster and more reliable velocity sensors.

Laser velocimetry plays an important role, as it offers many advantages in comparison to conventional anemometry methods: hotwire anemometry, Rumbo anemometry, ultrasound anemometry, etc. It provides contact-free, quick measurements at high resolution that can be conducted in harsh environments.

In this chapter, we have covered only the several approaches existing in modern velocimetry. We featured laser systems based on Doppler effect, such as lidars, microscopes, low-coherence tomographs, and on imaging modalities, and illustrated their performance, possibilities, and potentialities by the examples that we considered most interesting. Space limitations did not allow us to discuss velocimetry techniques based on laser speckle and fluorescence that find many important applications. Such interesting phenomena as self-mixing on which many velocity sensors are based was also left beyond the scope of this chapter. This can only be explained by the limited range of personal interests of the author.

However, the major conclusion is that new developments and applications in laser velocimetry are to appear because a variety of interesting, novel researches in this field are under way.

References

1. J. Workman, A. Springsteen (eds.), *Applied Spectroscopy A Compact Reference for Practioners* (Academic Press, San Diego, 1998)
2. J. Kauppinen, J. Partanen, *Fourier Transform Spectroscopy* (Wiley-VCH, Berlin, 2001)
3. D.M. Scott, A two-colour near-infrared sensor for sorting recycled plastic waste. Meas. Sci. Technol. **6**, 156–159 (1995)
4. V. Lucarini, J.J. Saarinen, K.-E. Peiponen, E.M. Vartiainen, *Kramers-Kronig Relations in Optical Materials Research* (Springer, Berlin, 2005)
5. A.L. Waterhouse, J.A. Kennedy, *Red Wine Color* (American Chemical Society, Washington, DC, 2004)
6. I. Noiseux, W. Long, A. Cournoyer, Simple fiber-optic-based sensor for process monitoring: An application in wine quality control monitoring. Appl. Spectrosc. **58**, 1010–1019 (2004)
7. C.F. Bohren, D.R. Huffman, *Absorption and Scattering of Light by Small Particles* (Wiley, New York, 1983)
8. L.M.C. Oliveira, M.A.C.P. Clemente, Port wine spectronephelometry. Opt. Laser Technol. **35**, 491–496 (2003)
9. J. Räty, K.-E. Peiponen, T. Asakura, *UV–Visible Reflection Spectroscopy of Liquids* (Springer, Berlin, 2004)
10. K.-E. Peiponen, E.M. Vartiainen, T. Asakura, *Dispersion, Complex Analysis and Optical Spectroscopy* (Springer, Berlin, 1999)
11. K.-E. Peiponen, E.M. Vartiainen, Dispersion theory of the reflectivity of s-polarized and p-polarized light. J. Opt. Soc. Am. B. **23**, 114–119 (2006)
12. M.O.A. Mäkinen, K.-E. Peiponen, J. Räty, V. Hyvärinen, Reflectance and probe window contamination: study of pulping solutions. Appl. Spectrosc. **55**, 852–857 (2001)
13. H. Soetedjo, J. Räty, Reflectometric study of contaminat layer on a probe window. Appl. Spectrosc. **57**, 915–919 (2003)
14. I. Niskanen, J. Räty, K.-E. Peiponen, Complex refractive index of tirbid liquids, Opt. Lett. **32**, 862–864 (2007)
15. I. Niskanen, J. Räty, K.-E. Peiponen, H. Koivula, M. Toivakka, Assessment of the complex refractive index of an optically very dense solid layer: Case study of offset magenta ink. Chem. Phys. Lett. **442**, 515–517 (2007)

16. I. Niskanen, J. Räty, K.-E. Peiponen, Estimation of effective refractive index of birifringent particles using a combination of the immersion liquid method and light scattering. Appl. Spectrosc. **62**, 399–401 (2008)

17. H. Räther, *Surface Plasmons on Smooth and Rough Surfaces and on Gratings* (Springer, Berlin, 1988)

18. J. Homola, S.S. Yee, G. Gauglitz, Surface plasmon resonance sensors: review. Sens. Actuators B **54**, 3–15 (1999)

19. K. Matsubara, S. Kawata, S. Minami, Optical chemical sensor based on surface plasmon measurement. Appl. Opt. **27**, 1160–1163 (1988)

20. R.J. Green, R.A. Frazier, K.M. Shakesheff, M.C. Davies, C.J. Roberts, S.J.B. Tendler, Surface plasmon resonance analysis of dynamic biological interactions with biomaterials. Biomaterials **21**, 1823–1835 (2000)

21. B.J. Sedlak, Next generation microarray technologies – focus is on higher sensitivity, drug discovery, and lipid cell signaling. Genet. Eng. News **23**, 20 (2003)

22. J. Liu, S. Tian, L. Tiefenauer, P.E. Nielsen, W. Knoll, Simultaneously amplified electrochemical and surface plasmon optical detection of DNA hybridiztion based on ferrocene-streptavidin conjugates. Anal. Chem. **77**, 2756–2761 (2005)

23. A.J. Jääskeläinen, K.-E. Peiponen, J.A. Räty, On reflectometric measurement of a refractive index of milk. J. Dairy Sci. **84**, 38–43 (2001)

24. M. Zangeneh, N. Doa, E. Sambriski, R.H. Terrill, Surface plasmon spectral finger-printing of adsorbed magnesium phthalocyanine by angle and wavelength modulation. Appl. Spectrosc. **58**, 10–17 (2004)

25. A.J. Haes, R.P. Van Duyne, Preliminary studies and potential applications of localized surface plasmon resonance spectroscopy in medical diagnostics. Expert Rev. Mol. Diagn. **4**, 527–537 (2004)

26. E.M. Vartiainen, J.J. Saarinen, K.-E. Peiponen, Method for extracting the complex dielectric function of nanospheres in a water matrix from surface plasmon resonance data. J. Opt. Soc. Am. B **22**, 1173–1178 (2005)

27. P. Kubelka, F. Munk, Ein Beitrag zur Optik der Farbanstriche. Zeitschrift für technische Physik **12**, 593–601 (1931)

28. G. Kortűm, *Reflectance Spectroscopy Principles, Methods, Applications* (Springer, Berlin, 1969)

29. D.A.G. Bruggeman, Berechnung versciedener physikalicher Konstanten von heterogenen Substanzen. Ann. Phys. (Leipzig) **24**, 636–679 (1935)

30. X.C. Zeng, D.J. Bergman, P.M. Hui, D. Stroud, Effective-medium theory for weakly nonlinear composites. Phys. Rev. B **38**, 10970–10973 (1988)

31. K.-E. Peiponen, E. Gornov, Description of Wiener bounds of multi-component composites by barycentric coordinates. Opt. Lett. (in press)

32. A.J. Jääskeläinen, K.-E. Peiponen, J. Räty, U. Tapper, O. Richard, E.I. Kauppinen, K. Lumme, Estimation of the refractive index of plastic pigments by Wiener bounds. Opt. Eng. **39**, 2959–2963 (2000)

33. K.-E. Peiponen, E. Gornov, On prediction of optical properties of two- and multiphase nanocomposites for nanomedicine. Int. J. Nanomed. **2**, 799–804 (2007)

34. Y.R. Shen, *The Principles of Nonlinear Optics* (Wiley, New York, 1984)

35. R.W. Boyd, *Nonlinear Optics* (Academic Press, New York, 2003)

36. P. Hänninen, A. Soini, N. Meltola, J. Soini, J. Soukka, E. Soini, A new microvolume technique for bioaffinity assays using two-photon excitation. Nat. Biotechnol. **18**, 548–550 (2000)

37. J. Bretschneider, Application of the optical testing procedure to quality control of flat glass. Glasstech. Ber. **61**, 172–175 (1988)

38. J. Räsänen, K.-E. Peiponen, On-line measurement of the thickness and optical quality of float glass with a sensor based on a diffractive element. Appl. Opt. **40**, 5034–5039 (2001)

39. B.R. Brown, A.W. Lohmann, Complex spatial filtering with binary masks. Appl. Opt. **5**, 967 (1966)

40. M. Nieto-Vesperinas, *Scattering and Diffraction in Physical Optics* (Wiley, New York, 1991)

41. S.M. Chapman, Pulp Paper Mag. Can. **55**, 88 (1954)

42. G. Blokhuis, P.J. Kalff, Tappi J. **59**, 107 (1976)

43. A. Oksman, R. Silvennoinen, K.-E. Peiponen, M. Avikainen, H. Komulainen, Reflectance study of paper. Appl. Spectrosc. **58**, 481–485 (2004)

44. M. Aikio, Hyperspectral prism-grating-prism imaging spectrograph. *VTT Publications* 435, PhD Dissertation (2001)

45. D.J. Whitehouse, A philosophy of linking manufacture to function-an example in optics. Proc. Inst. Mech. Engrs. **207**, 31–42 (1993)

46. J.M. Bennett, L. Matsson, *Introduction to Surface Roughness and Scattering* (Optical Society of America, Washington DC, 1989)

47. A. Ogilvy, J.R. Foster, Rough surfaces: gaussian or exponential statistics? J. Phys. D: Appl. Phys. **22**, 1243–1251 (1989)

48. P. Beckmann, A. Spizzcihino, *The Scattering of Electromagnetic Waves from Rough Surfaces* (Pergamon Press, Oxford, 1963)

49. P. Cielo, *Optical Techniques for Industrial Inspection* (Academic Press, San Diego, 1988)

50. H. Davies, The reflection of electromagnetic waves from rough surface, *Monograph No 90 RADIO SECTION*, 209–214 (1954)

51. R. Silvennoinen, K.-E. Peiponen, V. Hyvärinen, P. Raatikainen, P. Paronen, Optical surface roughness study of starch acetate compacts. Int. J. Pharm. **182**, 213–220 (1999)

52. V. Hyvärinen, K.-E. Peiponen, R. Silvennoinen, P. Raatikainen, P. Paronen, T. Niskanen, Optical inspection of punches: flat surfaces. Eur. J. Pharm. Biopharm. **49**, 87–90 (2000)

53. G.S. Spagnolo, D. Ambrosini, Diffractive optical element based sensor for roughness measurement. Sens. Actuators A **100**, 180–186 (2002)

54. R. Silvennoinen, K.-E. Peiponen, T. Asakura, Diffractive optical elements in materials inspection, in *International Trends in Optics and Photonics ICO IV*, ed. by T. Asakura (Springer, Berlin, 1999), pp. 281–293

55. C. Gu, K.-E. Peiponen, R. Silvennoinen, J. Luostarinen, J. Uozumi, T. Asakura, A simple proximity sensor for metal surface quality monitoring. Precis. Eng. **16**, 219–222 (1994)

56. K.-E. Peiponen, E. Alarousu, M. Juuti, R. Silvennoinen, A. Oksman, R. Myllylä, T. Prykäri, Diffractive-optical element-based glossmeter and low coherence interferometer in assessment of local surface quality of paper. Opt. Eng. **45**, 436011–436017 (2006)

57. M. Juuti, T. Prykäri, E. Alarousu, H. Koivula, M. Myllys, A. Lähteelä, M. Toivakka, J. Timonen, R. Myllylä, K.-E. Peiponen, Detection of local specular gloss and surface roughness from black prints. Colloids Surf. A **299**, 101–108 (2007)

58. V.G.W. Harrison, *Definition and Measurement of Gloss* (W. Hefner and Sons Ltd, Cambridge, 1940)

59. R.S. Hunter, R.W. Harold, *The Measurement of Appearance* (Wiley, New York, 1987)

60. J.S. Christie, An instrument for the geometric attibutes of metallic appearance. Appl. Opt. **8**, 1777–1785 (1969)

61. H. Assender, V. Bliznyuk, K. Porfyrakis, How surface topography relates to materials. Properties. Science **297**, 973–976 (2002)

62. T. Glatter, D.W. Bousfield, Print gloss development on a model substrate. Tappi J. **80** 125–131 (1997)

63. J.S. Preston, N.J. Elton, J.C. Husband, J. Dalton, P.J. Heard, G.C. Allen, Investigation into the distribution of ink components on printed coated paper. Part 1: optical and roughness considerations. Colloids Surf. A **205**, 183–198 (2002)

64. K. Myller, K.-E. Peiponen, R. Silvennoinen, J.-P. Tarvainen, J. Rainio, S. Soinila-Oksanen, Glossmeter for detection of gloss and wear of concave glazed ceramic products. cfi/Ber. DKG **81**, E39–E42 (2004)

65. K. Myller, M. Juuti, K.-E. Peiponen, R. Silvennoinen, E. Heikkinen, Quality inspection of metal surfaces by diffractive optical glossmeter. Precis. Eng. **30**, 443–447 (2006)

66. K.-E. Peiponen, M. Juuti, Statistical parameters for gloss evaluation. Appl. Phys. Lett. **88**, 0711041–0711043 (2006)

67. A. Oksman, M. Juuti, K.-E. Peiponen, Sensor for the detection of local contrast gloss of products. Opt. Lett. **33**, 654–656 (2008)

68. A. Oksman, M. Juuti, K.-E. Peiponen, Statistical parameters and analysis of local contrast gloss. Opt. Express **16**, 12415–12422 (2008)

69. A. Mäkynen, Position-sensitive devices and sensor systems for optical tracking and displacement sensing applications. Acta Univ Oul **C 151**, DSc Thesis (2000)

70. T. Bosch, M. Lescure (eds.), *Selected Papers on Laser Distance Measurement* (SPIE Milestone Series **MS 115**, 1995)

71. M.-C. Amann, T. Bosch, M. Lescure, R. Myllylä, M. Rioux, Laser ranging: a critical review of usual techniques for distance measurement. Opt. Eng. **40**, 10–19 (2001)

72. S. Donati, *Electro-Optical Instrumentation* (Prentice Hall, New Jersey, 2004)

73. T. Dresel, G. Häusler, H. Venzke, Three-dimensional sensing of rough surfaces by coherence radar. Appl. Opt. **31**, 919–925 (1992)

74. M.E. Brezinski, J.G. Fujimoto, Optical coherence tomography: High-resolution imaging in nontransparent tissue. IEEE J. Select. Top. Quantum Electron. **5**, 1185–1192 (1999)

75. P.H. Tomlins, R.K. Wang, Theory, developments and applications of optical coherence tomography. J. Phys. D: Appl. Phys. **38**, 2519–2535 (2005)

76. E. Alarousu, Low coherence interferometry and optical coherence tomography in paper measurements. Acta Univ. Oul **C 256**, DSc Thesis (2006)

77. S. Poujouly, B. Journet, Laser range finding by phase-shift measurement: moving towards smart systems. Proc. SPIE **4189**, 152–160 (2000)

78. B. Journet, G. Bazin, A low-cost laser range-finder based on an FMCW-like method. IEEE Trans. Instrum. Meas. **49**, 840–843 (2000)

79. I. Kaisto, J. Kostamovaara, M. Manninen, R. Myllylä, Optical range finder for 1.5–10 m distances. Appl. Opt. **22**, 3258–3264 (1983)

80. D. Castagnet, H. Tap-Béteille, M. Lescure, Avalanche-photodiode-based heterodyne optical head of a phase-shift laser range finder. Opt. Eng., **45**, 043003-1–043003-7 (2006)

81. C.E. Cook, M. Bernfield, *Radar Signals, An Introduction to Theory and Application* (Academic, New York, 1967)

82. N.J. Mohamed, Resolution function of nonsinusoidal radar signals: II- range-velocity resolution with pulse compression techniques. IEEE Trans. Electromagn. Compat. **33**, 51–58 (1991)

83. T. Ruotsalainen, P. Palojärvi, J. Kostamovaara, A current-mode gain-control scheme with constant bandwidth and propagation delay for transimpedance preamplifier. IEEE J. Solid-State Circuits **34**, 253–258 (1999)

84. J.-P. Jansson, A. Mäntyniemi, J. Kostamovaara, A CMOS time-to-digital converter with better than 10 ps single-shot precision. IEEE J. Solid-State Circuits **41**, 1286–1296 (2006)

85. K. Määttä, J. Kostamovaara, R. Myllylä, Profiling of hot surfaces by pulsed time-of-flight laser range finder techniques. Appl. Opt. **32**, 5334–5347 (1993)

86. M.R. Maier, P. Sperr, On the construction of a fast constant fraction trigger with integrated circuits and application to various phototubes. Nucl. Instrum. Methods **87**, 13–18 (1970)

87. A.J. Joblin, Method of calculating the image resolution of a near-infrared time-of-flight tissue-imaging system. Appl. Opt. **35**, 1996, 752–757 (1996)

88. J. Carlsson, P. Hellentin, A. Malmqvist, W. Persson, C-G Wahlström, Time-resolved studies of light propagation in paper. Appl. Opt. **34**, 1528–1535 (1995)

89. H. Lahtinen, M. Jurvakainen, P. Pramila, H. Tabell, M. Kusevic, O. Hormi, V. Lyöri, M. Heikkinen, R. Myllylä, P. Suopajärvi, H. Kopola, Utilisation of optical fibre measurement techniques in determination of residual stresses of composites. Proc. SPIE **3746**, 526–529 (1999)

90. A. Kilpelä, Pulsed time-of-flight laser ranger finder techniques for fast, high precision measurement applications. Acta Univ Oul **C 197**, DSc Thesis (2004)

91. R. Myllylä, J. Marszalec, J. Kostamovaara, A Mäntyniemi, G.-J. Ulbrich, Imaging distance measurements using TOF lidar. J. Opt. **29**, 188–193 (1998)

92. J. Varela, A PET imaging system dedicated to mammography. Radiat. Phys. Chem. 76, 347–350 (2007)

93. M.A. Albota, R.M. Heinrichs, D.G. Kocher, D.G. Fouche, B.E. Player, M.E. O'Brien, B.F. Aull, J.J. Zayhowski, J. Mooney, B.C. Willard, R.R. Carlson, Three-dimensional imaging laser radar with a photon-counting avalanche photodiode array and microchip laser. Appl. Opt. **41**, 7671–7678 (2002)

94. B.F. Aull, A.H. Loomis, D.J. Young, R.M. Heinrichs, B.J. Felton, P.J. Daniels, D.J. Landers, Geiger-Mode avalanche photodiodes for three-dimensional imaging. Lincoln Lab. J. **13**, 335–350 (2002)

95. J.W. Weingarten, G. Gruener, R. Siegwari, A state-of-the-art 3D sensor for robot navigation. Proc. 2004 IEEE/RSJ Int. Conf. Intell. Robots Syst., September 28. October 2, 2155–2160 (2004)

96. J. Busck, H. Heiselberg, Gated viewing and high-accuracy three-dimensional laser radar. Appl. Opt. **43**, 4705–4710 (2004)

97. P. Andersson, Long-range three-dimensional imaging using range-gated laser radar images. Opt. Eng. **45**, 034301-1-034301-10 (2006)

98. J.J. Zayhowski, A.L. Wilson Jr., Miniature eye-safe laser system for high-resolution three-dimensional lidar. Appl. Opt. **46**, 5951–5956 (2007)

99. A. Mäkynen, J. Kostamovaara, R. Myllyla, Displacement sensing resolution of position-sensitive detectors in atmospheric turbulence using retroreflected beam. IEEE Trans. Instrum. Meas. **46**, 1133–1136 (1997)

100. S. Donati, C.-Y. Chen, C.-C. Yang, Uncertainty of positioning and displacement measurements in quantum and thermal regimes. IEEE Trans. Instrum. Meas. **56** (2007), 1658–1665 (2007)

101. A. Mäkynen, J. Kostamovaara, R. Myllylä, Positioning resolution of the position-sensitive detectors in high background illumination. IEEE Trans. Instrum. Meas. **45**, 324–326 (1996)

102. A. Mäkynen, J. Kostamovaara, Accuracy of lateral displacement sensing in atmospheric turbulence using a retroreflector and a position-sensitive detector. Opt. Eng. **36**, 3119–3126 (1997)

103. J.H. Churnside, R.J. Lataitis, Wander of an optical beam in the turbulent atmosphere. Appl. Opt. **29**, 926–930 (1990)

104. M.L. Plett, Free-space optical communication link across 16 kilometers to a modulated retro-reflector array. Dissertation Submitted to the Faculty of the Graduate School of the University of Maryland, College Park, Electrical Engineering, 1–165 (2007)

105. A. Mäkynen, J. Kostamovaara, R. Myllylä, A high resolution lateral displacement sensing method using active illumination of a cooperative target and focused four- quadrant position-sensitive detector. IEEE Trans. Instrum. Meas. **44**, 46–52 (1995)

106. M. Tervaskanto, Laser targeted Finland focus. Traffic Technol. Int., Dec 2004/Jan 2005, 48–50 (2005)

107. D. Douxchamps, B. Macq, K. Chihara, High accuracy traffic monitoring using road-side line-scan cameras. Proc. IEEE ITSC 2006, 875–878 (2006)

108. V. Tuchin, *Tissue Optics* (SPIE Press, Bellingham, 2000)

109. B. Chance, Time resolved spectroscopic (TRS) and continuous wave (CWS) studies of photon migration in human arms and limbs. Adv. Exp. Med. Biol. **248**, 21–31 (1989)

110. T. Svensson, J. Swartling, P. Taroni, A. Torricelli, P. Lindblom, C. Ingvar, S. Andersson-Engels, Characterization of normal breast tissue heterogeneity using time-resolved near-infrared spectroscopy. Phys. Med. Biol. **50**, 2559–2571 (2005)

111. L.C. Enfield, A.P. Gibson, N.L. Everdell, D.T. Delpy, M. Schweiger, S.R. Arridge, C. Richardson, M. Keshtgar, M. Douek, J.C. Hebden, Three-dimensional time-resolved optical mammography of the uncompressed breast. Appl. Opt. **46**, 3628–3638 (2007)

112. V.V. Tuchin, X. Xu, R.K. Wang, Dynamic optical coherence tomography in studies of optical clearing, sedimentation, and aggregation of immersed blood. Appl. Opt. **41**, 258–271 (2002)

113. J. Saarela, Photon migration in pulp and paper. Acta Univ Oul **C 213**, DSc Thesis (2004)

114. A. Karppinen, A. Kilpelä, M. Karras, J. Tornberg, R. Myllylä, Papermaking furnish properties estimated by time-resolved spectroscopy. J. Pulp Paper Sci. **21**, J151–J154 (1995)

115. J. Saarela, R. Myllylä, Changes in the time-of-flight of a laser pulse during paper compression. J. Pulp Paper Sci. **29**, 224–227 (2003)

116. T. Fabritius, Optical method for liquid sorption measurements in paper. Acta Univ Oul **C 269**, DSc Thesis (2007)

117. D. Stifter, Beyond biomedicine: a review of alternative applications and developments for optical coherence tomography. Appl. Phys. **B88**, 337–357 (2007)
118. M. Jurvakainen, H. Lahtinen, P. Peltomäki, A. Pramila, V. Lyori, R. Myllyla, M. Heikkinen, P. Suopajarvi, H. Kopola, Determination of farfield strain in a composite structure using time-of-flight and fabry-perot optical fibre sensors. Proc. SPIE **4074**, 427–434 (2000)
119. T. Myllylä, H. Lahtinen, H. Sorvoja, R. Myllylä, Case studies on a wireless fibre Bragg grating condition monitoring system for rotating composite cylinders. Conf. Networked Sens. Syst. INSS'07, 69–72 (2007)
120. V. Lyöri, A. Mäntyniemi, A. Kilpelä, Q. Duan, J. Kostamovaara, A fibre-optic time-of-flight radar with a sub-metre spatial Resolution for the measurement of integral strain. Proc. SPIE **5050**, 322–332 (2003)
121. V. Lyöri, A. Kilpelä, G. Duan, A. Mäntyniemi, J. Kostamovaara, Pulsed time-of-flight radar for fiber-optic strain sensing. Rev. Sci. Instrum. **78**, 0247051–0247058 (2007)
122. B. Lichtberger, *Track Compendium* (Eurailpress, Hamburg, 2005)
123. J. Heinula, V. Nissinen, Shooting training: a principled approach. MS&T Mag. **2**, 36–38 (2003)
124. J.T. Viitasalo, P. Era, N. Konttinen, H. Mononen, K. Mononen, K. Norvapalo, Effects of 12-week shooting training and mode of feedback on shooting scores among novice shooters. Scand. J. Med. Sci. Sports **11**, 362–368 (2001)
125. Y. Yeh, H.Z. Cummins, Localized fluid flow measurements with an He–Ne laser spectrometer. Appl. Phys. Lett. **4**, 176 (1964)
126. H.E. Albrecht, M. Borys, N. Damaschke, C. Tropea, *Laser Doppler and Phase Doppler Measurement Techniques* (Springer, Berlin, 2003)
127. R.J. Adrian (ed.), *Selected Papers on Laser Doppler Velocimetry*, SPIE MS78 Belligham (1993)
128. R.J. Adrian (ed.), *Laser Techniques and Applications in Fluid Mechanics* (Springer, London, 1996)
129. Yu.N. Dubnistchev, B.S. Rinkevichyus (eds.), Optical methods of flow investigation. SPIE **6262** (2006)
130. H. Nobach, Analysis of dual-burst laser Doppler signals. Meas. Sci. Technol. **13**, 33–44 (2002)
131. L. Büttner, J. Czarske, Passive directional discrimination in Laser-Doppler Anemometry by the two-wavelength quadrature homodyne technique. Appl. Opt. **42**, 3843–3852 (2003)
132. P.J. Cronin, C.J. Cogswell, Minimum theoretical requirements for three-dimensional scanning-laser doppler anemometry. Appl. Opt. **39**, 6350–6359 (2000)
133. E.-J. Nijhof, W.S.J. Uijttewaal, R.M. Heethaar, Blood particle distributions accessed by microscopic laser Doppler velocimetry. Proc. SPIE **2052**, 187–194 (1993)
134. L. Büttner, J. Czarske, A multimode-fibre laser-Doppler anemometer for highly spatially resolved velocity measurements using low-coherence light. Meas. Sci. Technol. **12**, 1891–1903 (2001)
135. K. Shirai, T. Pfister, L. Büttner, J. Czarske, H. Müller, S. Becker, H. Lienhart, F. Durst, Highly spatially resolved velocity measurements of a turbulent channel flow by a fiber-optic heterodyne laser-Doppler velocity-profile sensor. Exp. Fluids **40**, 473–481 (2006)

136. T. Pfister, L. Büttner, K. Shirai, J. Czarske, Monochromatic heterodyne fiber-optic profile sensor for spatially resolved velocity measurements with frequency division multiplexing. Appl. Opt. **44**, 2501–2510 (2005)

137. K. Shirai, T. Pfister, L. Büttner, J. Czarske, H. Müller, S. Becker, H. Lienhart, F. Durst, Highly spatially resolved velocity measurements of a turbulent channel flow by a fiber-optic heterodyne laser-Doppler velocity-profile sensor. Exp. Fluids **40**, 473–481 (2006)

138. V.M. Gordienko, A.A Kormakov, L.A. Kosovsky, N.N. Kurochkin, G.A. Pogosov, A.V. Priezzhev, Yu. Ya Putivskii, Coherent CO_2 lidars for measuring wind velocity and atmospheric turbulence. Opt. Eng. **33**(10), 3206–3213 (1994)

139. C. Werner, Applications of space-borne Doppler and backscatter lidar: a critical review. Opt. Eng. **34**(11), 3103–3114 (1995)

140. S. Rahm, Measurement of a wind field with an air-borne continuous-wave Doppler lidar. Opt. Lett. **20**, 216–218 (1995)

141. E. Galletti, E. Stucchi, D.V. Willetts, M.R. Harris, Transverse-mode selection in apertured super-Gaussian resonators: an experimental and numerical investigation for a pulsed CO_2 Doppler lidar transmitter. Appl. Opt. **36**, 1269–1277 (1997)

142. T.J. Kane, W.J. Kozlovsky, R.L. Byer, C.E. Byvik, Coherent laser radar at 1.06 μm using Nd:YAG lasers. Opt. Lett. **12**, 239–241 (1987)

143. V.M. Gordienko, A.N. Konovalov, N.V. Kravtsov, Yu. Ya Putivskii, V.I. Savin, V.V. Firsov, Remote Doppler velocimeter based on an Nd^{3+}:YAG chip laser and its application in a study of laser-induced hydrodynamic flows. Quant. Electron. **28**(9), 827–830 (1998)

144. K. Otsuka, Ultrahigh sensitivity laser Doppler velocimetry with a microchip solid-state laser. Appl. Opt. **33**, 1111–1114 (1994)

145. C.J. Karlsson, F. Å.A. Olsson, D. Letalick, M. Harris, All-fiber multifunction continuous-wave coherent laser radar at 1.55 μm for range, speed, vibration, and wind measurements. Appl. Opt. **39**, 3716–3726 (2000)

146. M. Harris, G.N. Pearson, K.D. Ridley, C.J. Karlsson, F. Å.A. Olsson, D. Letalick, Single-particle laser Doppler anemometry at 1.55 μm. Appl. Opt. **40**, 969–973 (2001)

147. M. Harris, G. Constant, C. Ward, Continuous-wave bistatic laser Doppler wind sensor. Appl. Opt. **40**, 1501–1506 (2001)

148. G.J. Koch, J.Y. Beyon, B.W. Barnes, M. Petros, J. Yu, F. Amzajerdian, M.J. Kavaya, U.N. Singh, High-energy 2 μm Doppler lidar for wind measurements. Opt. Eng. **46**(11), 116201 (2007)

149. R.V. Mustasich, B.R. Ware, A study of protoplasmic streaming in Nitella by laser Doppler spectroscopy. Biophys. J. **16**, 373–388 (1976)

150. R.V. Mustasich, D.B. Sattelle, P.B. Buchan, Cytoplasmic streaming in Chara corallina studied by laser light scattering. J. Cell. Sci. **22**(3), 633–643 (1976)

151. D. Ackers, Z. Hejnowicz, A. Sievers, Variation in velocity of cytoplasmic streaming and gravity effect in characean internodal cells measured by laser-Doppler-velocimetry. Protoplasma **179**(1–2), 61–71 (1994)

152. D. Ackers, B. Buchen, Z. Hejnowicz, A. Sievers, The pattern of acropetal and basipetal cytoplasmic streaming velocities in Chara rhizoids and protonemata, and gravity effect on the pattern as measured by laser-Doppler-velocimetry. Planta **211**(1), 133–143 (2000)

153. Yu.A. Denisov, A.S. Stepanian, A.V. Priezzhev, Laser Doppler spectroscopy of blood flow. Moscow Univ. Phys. Bull. Ser. 3, **30**, 62–66 (1989)

154. A.V. Priezzhev, A.S. Stepanian, Peculiarities of blood flow velocity measurement in thin capillaries and possibilities of laser methods. Laser Med. **1**, 31–34 (1997)

155. S.E. Skipetrov, R. Maynard, Dynamic multiple scattering of light in multilayer turbid media. Phys. Lett. A **217**, 181–185 (1996)

156. S.E. Skipetrov, I.V. Meglinsky, Wave-diffusion spectroscopy in random-inhomogenious media with local streams of scattering particles. J. Exp. Theor. Phys. **113**, 1213–1222 (1998)

157. A.V. Priezzhev, M.S. Polyakova, K.B. Begun, K. Vanag, A.F. Pogrebnaya, Dual-beam laser Doppler microscopy of suspension flows embedded into medium with strong scattering. Proc. SPIE **3915**, 129–136 (2000)

158. X.J. Wang, T.E. Milner, J.S. Nelson, Characterization of fluid flow velocity by optical Doppler tomography. Opt. Lett. **20**, 1337–1339 (1995)

159. J.A. Izatt, M.D. Kulkarni, S. Yazdanfar, J.K. Barton, A.J. Welch, In vivo bidirectional color Doppler flow imaging of picoliter blood volumes using optical coherence tomography. Opt. Lett. **22**, 1439–1441 (1997)

160. Z. Chen, T.E. Milner, D. Dave, J.S. Nelson, Optical Doppler tomography imaging of fluid flow velocity in highly scattering media. Opt. Lett. **22**, 64–66 (1997)

161. Y. Zhao, Z. Chen, C. Saxer, S. Xiang, J.F. de Boer, J.S. Nelson, Phase resolved optical coherence tomography and optical Doppler tomography for imaging blood flow in human skin with fast scanning speed and high velocity sensitivity. Opt. Lett. **25**, 114–116 (2000)

162. J. Hast, T. Prykäri, E. Alarousu, R. Myllylä, A.V. Priezzhev, Flow imaging and velocity measurement of highly scattering liquid inside scattering media using Doppler optical coherence tomography. Proc. SPIE **4965**, 66–72 (2003)

163. X.J. Wang, T.E. Milner, Z. Chen, J.S. Nelson, Measurement of fluid-flow-velocity profile in turbid media by the use of optical Doppler tomography. Appl. Opt. **36**(1), 144–149 (1997)

164. S.G. Proskurin, Y. He, R.K. Wang, Doppler optical coherence imaging of converging flow. Phys. Med. Biol. **49**(7), 1265–1276 (2004)

165. M. Bonesi, D. Churmakov, I. Meglinski, Study of flow dynamics in complex vessels using Doppler optical coherence tomography. Meas. Sci. Technol. **18**, 3279–3286 (2007)

166. S.G. Proskurin, Y. He, R.K. Wang, Determination of flow velocity vector based on Doppler shift and spectrum broadening with optical coherence tomography. Opt. Lett. **28**, 1227–1229 (2003)

167. R.K. Wang, High-resolution visualization of fluid dynamics with Doppler optical coherence tomography. Meas. Sci. Technol. **15**, 725–733 (2004)

168. T. Lindmo, D.J. Smithies, Z. Chen, J.S. Nelson, T.E. Milner, Accuracy and noise in optical Doppler tomography studied by Monte Carlo simulation. Phys. Med. Biol. **43**, 3045–3064 (1998)

169. A.V. Bykov, M.Yu. Kirillin, A.V. Priezzhev, Monte Carlo simulation of an optical coherence Doppler tomograph signal: the effect of the concentration of particles in a flow on the reconstructed velocity profile. Quant. Electron. **35**(2), 135–139 (2005)

170. A.V. Bykov, M. Yu. Kirillin, A.V. Priezzhev, Analysis of distortions in the velocity profiles of suspension flows in a light-scattering medium upon their reconstruction from the optical coherence Doppler tomograph signal. Quant. Electron. **35**(11), 1079–1082 (2005)

171. J. Moger, S.J. Matcher, C.P. Winlove, A. Shore, The effect of multiple scattering on velocity profiles measured using Doppler OCT. J. Phys. D: Appl. Phys. **38**, 2597–2605 (2005)
172. J. Seki, Y. Satomura, Y. Ooi, T. Yanagida, A. Seiyama, Velocity profiles in the rat cerebral microvessels measured by optical coherence tomography. Clin. Hemorheol. Microcirc. **34**(1–2), 233–239 (2006)
173. S. Yazdanfa, A.M. Rollins, J.A. Izatt, In vivo imaging of human retinal flow dynamics by color Doppler optical coherence tomography. Arch. Ophthalmol. **121**, 235–239 (2003)
174. B.R. White, M.C. Pierce, N. Nassif, B. Cense, B.H. Park, G.J. Tearney, B. Bouma, T.C. Chen, J.F. de Boer, In vivo dynamic human retinal blood flow imaging using ultra-high-speed spectral domain optical Doppler tomography. Opt. Express **11**, 3490–3497 (2003)
175. M. Wojtkowski, V.J. Srinivasan, T.H. Ko, J.G. Fujimoto, A. Kowalczyk, J.S. Duker, Ultrahigh-resolution, high-speed, Fourier domain optical coherence tomography and methods for dispersion compensation. Opt. Express **12**, 2404–2422 (2004)
176. Y. Zhao, Z. Chen, C. Saxer, Q. Shen, S. Xiang, J.F. de Boer, J.S. Nelson, Doppler standard deviation imaging for clinical monitoring of in vivo human skin blood flow. Opt. Lett. **25**, 1358–1360 (2000)
177. Y. Wang, B.A. Bower, J.A. Izatt, O. Tan, D. Huang, In vivo total retinal blood flow measurement by Fourier domain Doppler optical coherence tomography. J. Biomed. Opt. **12**(4), 041215 (2007)
178. H. Ren, T. Sun, D.J. MacDonald, M.J. Cobb, X. Li, Real-time in vivo blood-flow imaging by moving-scatterer-sensitive spectral-domain optical Doppler tomography. Opt. Lett. **31**(7), 927–929 (2006)
179. V.X. Yang, Y.X. Mao, N. Munce, B. Standish, W. Kucharczyk, N.E. Marcon, B.C. Wilson, A.I. Vitkin, Interstitial Doppler optical coherence tomography. Opt Lett. **30**(14), 1791–1793 (2005)
180. H. Li, B.A. Standish, A. Mariampillai, N.R. Munce, Y. Mao, S. Chiu, N.E. Marcon, B.C. Wilson, A.I. Vitkin, V.X. Yang, Feasibility of interstitial Doppler optical coherence tomography for in vivo detection of microvascular changes during photodynamic therapy. Lasers Surg. Med. **38**(8), 754–761 (2006)
181. B.A. Standish, X. Jin, J.A. Smolen, A. Mariampillai, N.R. Munce, B.C. Wilson, A.I. Vitkin, V.X. Yang, Interstitial Doppler optical coherence tomography monitors microvascular changes during photodynamic therapy in a Dunning prostate model under varying treatment conditions. J. Biomed. Opt. **12**(03), 034022 (2007)
182. V.X.D. Yang, Y. Mao, B.A. Standish, N.R. Munce, S. Chiu, D. Burnes, B.C. Wilson, I.A. Vitkin, P.A. Himmer, D.L. Dickensheets, Doppler optical coherence tomography with a micro-electro-mechanical membrane mirror for high-speed dynamic focus tracking. Opt. Lett. **31**, 1262–1264 (2006)
183. V.X. Yang, S.J. Tang, M.L. Gordon, B. Qi, G. Gardiner, M. Cirocco, P. Kortan, G.B. Haber, G. Kandel, A.I. Vitkin, B.C. Wilson, N.E. Marcon, Endoscopic Doppler optical coherence tomography in the human GI tract: initial experience. Gastrointest. Endosc. **61**(7), 879–890 (2005)
184. M.D. Kulkarni, T.G. van Leeuwen, S. Yazdanfar, J.A. Izatt, Velocity-estimation accuracy and frame-rate limitations in color Doppler optical coherence tomography. Opt. Lett. **23**(13), 1057–1059 (1998)

185. M.A. Choma, A.K. Ellerbee, S. Yazdanfar, J.A. Izatt, Doppler flow imaging of cytoplasmic streaming using spectral domain phase microscopy. J. Biomed. Opt. **11**(2), 024014 (2006)

186. A.K. Ellerbee, H.C. Hendargo, A.R. Motomura, J.A. Izatt, Extension of spectral domain phase microscopy to three-dimensional nanoscale displacement mapping in cardiomyocytes. Proc. SPIE **6861**, 686108 (2008)

187. A.P. Shepherd, P.A. Öberg (eds.), *Laser-Doppler Blood Flowmetry* (Kluwer, Hingham, MA, 1990)

188. W. Steenbergen, R. Kolkman, F. de Mul, Light-scattering properties of undiluted human blood subjected to simple shear. J. Opt. Soc. Am. A **16**, 2959–2967 (1999)

189. A.N. Yaroslavsky, A.V. Priezzhev, J. Rodrigues, I.V. Yaroslavsky, H. Battarbee, Optics of blood, in *Handbook on Optical Biomedical Diagnostics*, ed. By V.V. Tuchin (SPIE Press, Bellingham, 2002), pp. 169–216

190. F.F.M. de Mul, M.H. Koelink, M.L. Kok, P.J. Harmsma, J. Greve, R. Graaff, J.G. Aarnoudse, Laser Doppler velocimetry and Monte Carlo simulations on models for blood perfusion in tissue. Appl. Opt. **34**, 6595–6611 (1995)

191. Y. Watanabe, E. Okada, Influence of perfusion depth on laser Doppler flow measurements with large source-detector spacing. Appl. Opt. **42**, 3198–3204 (2003)

192. M. Bracic, A. Stefanovska, Wavelet-based analysis of human blood-flow dynamics. Bull. Math. Biol. **60**(5), 919–935 (1998)

193. A. Stefanovska, M. Bracic, H.D. Kvernmo, Wavelet analysis of oscillations in the peripheral blood circulation measured by laser Doppler technique. IEEE Trans. Biomed. Eng. **46**(10), 1230–1239 (1999)

194. A. Serov, B. Steinacher, T. Lasser, Full-field laser Doppler blood-flow imaging and monitoring using an intelligent CMOS camera and area illumination. Opt. Express **13**(10), 3681–3689 (2005)

195. A. Serov, T. Lasser, High-speed laser Doppler perfusion imaging using an integrating CMOS image sensor. Opt. Express **13**(17), 6416–6428 (2005)

196. E.R. Fossum, CMOS image sensors: electronic camera-on-a-chip. IEEE Trans. Electron. Devices **44**, 1698–1698 (1997)

197. V.G. Kolinko, F.F.M. de Mul, J. Greve, A.V. Priezzhev, Feasibility of picosecond laser-Doppler flowmetry provides basis for time-resolved tomography of biological tissue. J. Biomed. Opt. **3**(2), 187–190 (1998)

198. S. Monstrey, H. Hoeksema, J. Verbelen, A. Pirayesh, P. Blondeel, Assessment of burn depth and burn wound healing potential. Burns **34**(6), 761–769 (2008)

199. M. Atlan, B.C. Forget, A.C. Boccara, T. Vitalis, A. Rancillac, A.K. Dunn, M. Gross, Cortical blood flow assessment with frequency-domain laser Doppler microscopy. J. Biomed. Opt. **12**(2), 024019 (2007)

200. R.K. Wang, S. Hurst, Mapping of cerebro-vascular blood perfusion in mice with skin and skull intact by Optical Micro-AngioGraphy at 1.3 μm wavelength. Opt. Express **15**(18), 11402–11412 (2007)

201. R.K. Wang, Directional blood flow imaging in volumetric optical microangiography achieved by digital frequency modulation. Opt. Lett. **33**(16), 1878–1880 (2008)

202. R.K. Wang, S.L. Jacques, Z. Ma, S. Hurst, S.R. Hanson, A. Gruber, Three dimensional optical angiography. Opt. Express **15**(7), 4083–4097 (2007)

203. R.K. Wang, Three dimensional optical angiography maps directional blood perfusion deep within microcirculation tissue beds in vivo. Phys. Med. Biol. **52**, N531–N537 (2007)

204. L. An, R.K. Wang, In vivo volumetric imaging of vascular perfusion within human retina and choroids with optical micro-angiography. Opt. Express **16**(15), 11438–11452 (2008)

205. T.J. Mueller, The 11th International symposium on flow visualization. J. Visual. **8**(2), 187–191 (2005)

206. J. Kompenhans, The 12th International symposium on flow visualization. J. Visual. **10**(1), 123–128 (2007)

207. I. Grant, Electronic proceedings editor, Proceedings of ISFV-12 (2006), Optimage Ltd., Ediburgh, UK, CD ROM

208. I. Grant (ed.), *Selected Papers on Particle Image Velocimetry*, SPIE **MS99**, Bellingham (1994)

209. J. Westerweel, Fundamentals of digital particle image velocimetry. Meas. Sci. Technol. **8**, 1379–1392 (1997)

210. A.J. Prasad, Particle image velocimetry. Current Sci. **79**, 51–57 (2000)

211. R.J. Adrian, Twenty years of particle image velocimetry, in *Proceedings of 12th International Symposium on Applications of Laser Techniques to Fluid Mechanics*, Lisbon, July 12–15 (2004)

212. K.S. Breuer (ed.), *Microscale Diagnostic Techniques* (Springer, New-York, 2005)

213. M. Raffel, C.E. Willert, S.T. Wereley, J. Kompenhans, *Particle Image Velocimetry: A Practical Guide*, 2nd ed. (Springer, Berlin, 2007)

214. A. Melling, Tracer particles and seeding for particle image velocimetry. Meas. Sci. Technol. **8**, 1406–1416 (1997)

215. A.J. Prasad, Particle image velocimetry. Current Sci. **79**, 51–57 (2000)

216. R. Meynart, *Particle Image Displacement Velocimetry* (1980), part of the Von Karman Institute for Fluid Dynamics lecture series, March 21–25 (1988)

217. P.J. Bryanston-Cross, T.R. Judge, C. Quan, G. Pugh, N. Corby, The application of digital particle image velocimetry (DPIV) to transonic flows. Prog. Aerospace Sci. **31**, 273–290 (1995)

218. C.M. Zettner, M. Yoda, A novel interfacial velocimetry technique with submicron spatial resolution. *54th Annual Meeting of the Division of Fluid Dynamics of American Physical Society*, November 18–20, San Diego, CA, USA (2001)

219. M.L. Smith, D.S. Long, E.R. Damiano, K. Ley, Near-wall micro-PIV reveals a hydrodynamically relevant endothelial surface layer in venules in vivo. Biophys. J. **85**(1), 637–645 (2003)

220. J.Y. Lee, H.S. Ji, S.J. Lee, Micro-PIV measurements of blood flow in extraembryonic blood vessels of chicken embryos. Physiol. Meas. **28**(10), 1149–1162 (2007)

221. Y. Sugii, S. Nishio, K. Okamoto, Measurement of a velocity field in microvessels using a high resolution PIV technique. Ann. N. Y. Acad. Sci. **972**, 331–336 (2002)

222. A. Nakano, Y. Sugii, M. Minamiyama, H. Niimi, Measurement of red cell velocity in microvessels using particle image velocimetry (PIV). Clin. Hemorheol. Microcirc. **29**(3–4), 445–455 (2003)

223. A.M. Fallon, L.P. Dasi, U.M. Marzec, S.R. Hanson, A.P. Yoganathan, Procoagulant properties of flow fields in stenotic and expansive orifices. Ann. Biomed. Eng. **36**(1), 1–13 (2008)

224. R. Kaminsky, U. Morbiducci, M. Rossi, L. Scalise, P. Verdonck, M. Grigioni, Time-resolved PIV technique for high temporal resolution measurement of mechanical prosthetic aortic valve fluid dynamics. J. Artif. Organs **30**(2), 153–162 (2007)

225. H.B. Kim, J. Hertzberg, C. Lanning, R. Shandas, Noninvasive measurement of steady and pulsating velocity profiles and shear rates in arteries using echo PIV: in vitro validation studies. Ann. Biomed. Eng. **32**(8), 1067–1076 (2004)

226. D. Sinton, Microscale flow visualization. Microfluid. Nanofluid. **1**(1), 2–21 (2004)

227. R. Lima, S. Wada, M. Takeda, K. Tsubota, T. Yamaguchi, In vitro confocal micro-PIV measurements of blood flow in a square microchannel: the effect of the haematocrit on instantaneous velocity profiles. J. Biomech. **40**(12), 2752–2757 (2007)

228. R. Lima, S. Wada, S. Tanaka, M. Takeda, T. Ishikawa, K. Tsubota, Y. Imai, T. Yamaguchi, In vitro blood flow in a rectangular PDMS microchannel: experimental observations using a confocal micro-PIV system. Biomed. Microdev. **10**(2), 153–167 (2008)

229. H. Kinoshita, S. Kaneda, T. Fujii, M. Oshima, Three-dimensional measurement and visualization of internal flow of a moving droplet using confocal micro-PIV. Lab Chip. **7**(3), 338–346 (2007)

230. N. Nève, J.K. Lingwood, J. Zimmerman, S.S. Kohles, D.C. Tretheway, The μ PIVOT: an integrated particle image velocimeter and optical tweezers instrument for microenvironment investigations. Meas. Sci. Technol. **19** 095403 (2008)

231. K.D. Hinsch, Holographic particle image velocimetry. Meas. Sci. Technol. **13**(7) R61 (2002)

232. K.D. Hinsch, S.F. Herrmann, Holographic particle image velocimetry. Meas. Sci. Technol. **15**(4) (2004)

233. H. Meng, G. Pan, Y. Pu, S.H. Woodward, Holographic particle image velocimetry: from film to digital recording. Meas. Sci. Technol. **15**(4), 673–685 (2004)

234. E. Malkiel, J.N. Abras, J. Katz, Automated scanning and measurements of particle distributions within a holographic reconstructed volume. Meas. Sci. Technol. **15**(4), 601–612 (2004)

235. S.F. Herrmann, K.D. Hinsch, Light-in-flight holographic particle image velocimetry for wind-tunnel applications. Meas. Sci. Technol. **15**(4), 613–621 (2004)

236. R. Alcock, C.P. Garner, N.A. Halliwell, J.M. Coupland, An enhanced HPIV configuration for flow measurement through thick distorting windows. Meas. Sci. Technol. **15**(4), 631–638 (2004)

237. D. Barnhart, W.D. Koek, T. Juchem, N. Hampp, J.M. Coupland, N.A. Halliwell, Bacteriorhodopsin as a high-resolution, high-capacity buffer for digital holographic measurements. Meas. Sci. Technol. **15**(4), 639–646 (2004)

238. N. Masuda, T. Ito, K. Kayama, H. Kono, S. Satake, T. Kunugi, K. Sato, Special purpose computer for digital holographic particle tracking velocimetry. Opt. Express **14**, 587–592 (2006)

239. K. Matsumoto, S. Takagi, T. Nakagaki, Locomotive mechanism of *Physarum* plasmodia based on spatiotemporal analysis of protoplasmic streaming. Biophys. J. **94**, 2492–2504 (2008)

240. S.A. Japee, C.G. Ellis, R.N. Pittman, Flow visualization tools for image analysis of capillary networks. Microcirc. **11**(1), 39–54 (2004)

241. M. Cutolo, C. Pizzorni, A. Sulli, Nailfold video-capillaroscopy in systemic scle-rosis. Z. Rheumatol. **63**(6), 457–462 (2004)

242. Yu. Gurfinkel, Computer capillaroscopy as a channel of local visualization, noninvasive diagnostics, and screening of substances in circulating blood. Proc. SPIE **4241**, 467–472 (2001)

243. Yu.I. Gurfinkel, V.M. Mikhailov, New potentialities for noninvasive optical investigation of microcirculation in extended space missions. Proc. SPIE **4624**, 134–138 (2002)

244. Yu.I. Gurfinkel, V.M. Mikhailov, M.I. Kudutkina, Noninvasive estimation of tissue edema in healthy volunteers and in patients suffering from heart failure. Proc. SPIE **5325**, 150–156 (2004)

245. A.A. Parthasarathi, S.A. Japee, R.N. Pittman, Determination of red blood cell velocity by video shuttering and image analysis. Ann. Biomed. Eng. **27**(3), 313–325 (1999)

246. E. Chaigneau, M. Oheim, E. Audinat, S. Charpak, Two-photon imaging of capillary blood flow in olfactory bulb glomeruli. Proc. Natl Acad. Sci. Am. **100**(22), 13081–13086 (2003)

247. K. Tsukada, H. Minamitani, E. Sekizuka, C. Oshio, Image correlation method for measuring blood flow velocity in microcirculation: correlation 'window' sim-ulation and in vivo image analysis. Physiol. Meas. **21**, 459–471 (2000)

248. G. Mchedlishvili, Disturbed blood flow structuring as critical factor of hemorhe-ological disorders in microcirculation. Clin Hemorheol. Microcirc. **19**(4), 315–325 (1998)

Index

Springer Series in
OPTICAL SCIENCES

Springer Series in
OPTICAL SCIENCES

Lightning Source UK Ltd.
Milton Keynes UK
UKOW07n2113080215

245862UK00014B/200/P